META-LEVEL INFERENCE

STUDIES IN COMPUTER SCIENCE AND ARTIFICIAL INTELLIGENCE

1

Editors:

D. G. Bobrow
Xerox Corporation
Palo Alto Research Centre
Palo Alto, California

H. Kobayashi
IBM Japan Ltd.
Tokyo

J. Nievergelt
ETH, Institut für Informatik
Zürich

M. Nivat
Université Paris VII
Paris

NORTH-HOLLAND – AMSTERDAM • NEW YORK • OXFORD

META-LEVEL INFERENCE
*Representing and Learning Control Information
in Artificial Intelligence*

Bernard SILVER
*Fundamental Research Laboratory
GTE Laboratories Inc.*

1986

NORTH-HOLLAND – AMSTERDAM • NEW YORK • OXFORD

© Elsevier Science Publishers B.V., 1986

All rights reserved. No part of this publication may be reproduced, stored in a retrieval system, or transmitted, in any form or by any means, electronic, mechanical, photocopying, recording or otherwise, without the prior permission of the copyright owner.

ISBN: 0 444 87900 5

Publishers:
ELSEVIER SCIENCE PUBLISHERS B.V.
P.O. BOX 1991
1000 BZ AMSTERDAM
THE NETHERLANDS

Sole distributors for the U.S.A. and Canada:
ELSEVIER SCIENCE PUBLISHING COMPANY, INC.
52 VANDERBILT AVENUE
NEW YORK, N.Y. 10017
U.S.A.

Library of Congress Cataloging-in-Publication Data

Silver, Bernard, 1958-
  Meta-level inference.
  (Studies in computer science and artificial intelligence ; 1)

  Includes index.
  1. Artificial intelligence — Data processing. 2. Control theory.
3. Information theory. I. Title. II. Series
Q336.S55 1986   006.3   85-25445
ISBN 0-444-87900-5 (U.S.)

PRINTED IN THE NETHERLANDS

TO MARLENE

PREFACE

This book addresses two questions:

- How can search be controlled in domains with a large search space?

- How can this control information be learned?

It is argued that both problems can be tackled with the aid of a technique called **meta-level inference**. Two programs are presented in this book to support this claim.

The first program, PRESS[1], described in chapter 2, solves symbolic, transcendental, non-differential equations. The second program, LP[2], described in chapter 4 is also an equation solving program, but, unlike PRESS, it is capable of learning new equation-solving techniques. It embodies a new learning method, called **Precondition Analysis**. Precondition Analysis combines meta-level inference with concepts from the field of planning.

This book is a modified and extended version of the author's University of Edinburgh PhD thesis, [69].

Acknowledgements

I would like to express my gratitude to the following people:

To my supervisor, Alan Bundy, who has been of great help in many ways. He first interested me in the application of Artificial Intelligence to mathematical reasoning, in particular those aspects relating to equation solving and learning, and through many useful discussions he has helped me solve the problems I encountered. I would also like to thank him for his comments on drafts of this book.

Leon Sterling, my assistant supervisor, who has been similarly helpful in guiding my research. He has also provided many useful comments on drafts of this book.

My PhD examiners, Gordon Plotkin (Edinburgh) and Richard Young (Cambridge) for their suggestions for improving the presentation.

Richard O'Keefe, Lawrence Byrd, Lincoln Wallen, Jane Hesketh and Maarten van Sommeron, and other members of the PRESS project, who have provided help in many ways.

Marlene Kliman, for her comments on drafts of this book, for her proof-reading and for her help in many other ways.

---

[1] **P**rolog **E**quation **S**olving **S**ystem.

[2] **L**earning **P**RESS.

I have also had useful discussions with the following people: Pat Langley, Tom Mitchell, Jack Mostow, Ross Overbeek, Richard Waldinger and Ron Wenger.

This work was supported by the Science and Engineering Research Council, who provided grant GR/C/20826, and a Research Studentship for the author.

I would also like to thank GTE Laboratories for giving me the time and resources to work on the final version of this book.

TABLE OF CONTENTS

1. Introduction and Overview	1
1.1 The Problems	1
1.1.1 Search	1
1.1.2 Learning	2
1.2 Meta-level Inference	2
1.3 Precondition Analysis	4
1.3.1 The Operators of LP	4
1.4 The Choice of Domain	6
1.4.1 Equation Solving and Search	6
1.4.2 A level questions	7
1.5 Objectives and Motivations	8
1.5.1 The Objectives of the PRESS program	8
1.5.2 The Objectives of LP	9
1.5.2.1 The Self-Improving Algebra System	10
1.6 The Relevance of this work to Other Domains	10
1.7 The Structure of this book	11
1.7.1 Notation Conventions	11
2. Automated Equation Solving	13
2.1 An overview of PRESS	13
2.1.1 The Solution of an Equation	13
2.1.2 Solving Equations	14
2.2 Meta-level Inference in Equation Solving	14
2.2.1 The Exhaustive Application of Rewrite Rules	14
2.2.2 The Selective Application of Rewrite Rules	15
2.3 The Equation Solving Methods	16
2.3.1 The Individual Methods	18
2.3.1.1 Isolation	18
2.3.1.2 Collection	19
2.3.1.3 Attraction	21
2.3.1.4 The Role of the Worked Example	22
2.3.1.5 Factorization	22
2.3.1.6 Polynomial Solution Methods	23
2.3.1.7 Change of Unknown	24
2.3.1.8 Factorization Preparation	26
2.3.1.9 Trigonometric Factorization	27
2.3.1.10 Other Methods	29
2.3.2 Primary and Secondary Methods	29
2.3.3 The Heuristic Waterfall	29
2.3.3.1 Pruning the Meta-Level Search Space	30
2.3.4 Example Equations	31
2.4 Summary of the Methods of PRESS	32
2.5 Discussion	32

2.5.1 Discussion of Meta-level inference in PRESS ... 32
2.5.2 The Search Space ... 34
2.5.2.1 The Partitioning of the Rewrite Rules ... 35
2.5.2.2 The Control Information ... 36
2.5.3 Adding New Methods ... 36
2.5.3.1 The Performance of PRESS ... 37
2.6 MACSYMA ... 37
2.7 The Psychological Validity of PRESS ... 38
2.8 Conclusions ... 39
2.8.1 PRESS and LP ... 39

3. Homogenization ... 41

3.1 Standard Homogenization ... 41
 3.1.1 Some Terminology ... 42
 3.1.2 The Homogenization Method ... 42
 3.1.3 Selecting the Reduced Term ... 44
  3.1.3.1 The Degree of the Output Equation ... 44
  3.1.3.2 The Types of the Offenders Set ... 45
  3.1.3.3 The Simplicity Method ... 46
  3.1.3.4 The Mixed Offenders Set ... 46
  3.1.3.5 The Trigonometric Offenders Set ... 46
 3.1.4 Producing the Output Equation ... 48
 3.1.5 The Search Involved in Homogenization ... 49
  3.1.5.1 Results and example equations ... 49
  3.1.5.2 Performance of the Simplicity Method ... 49
 3.1.6 Cases when Homogenization should not be used ... 51
 3.1.7 Limitations and Possible Extensions ... 52
 3.1.8 Conclusions for Standard Homogenization ... 53
  3.1.8.1 The Need for Specialists ... 53
3.2 Extended Homogenization ... 54
 3.2.1 The Elementary Method ... 54
 3.2.2 Application of Homogenization to Simultaneous Equations ... 56
 3.2.3 Elimination ... 58
  3.2.3.1 The Method Of Elimination ... 59
 3.2.4 Conclusions for Extended Homogenization ... 61
3.3 Other Uses of Homogenization ... 62
 3.3.1 Homogenization in Unification ... 62
 3.3.2 Generalization ... 63
3.4 Conclusions ... 64
 3.4.1 Humans and Homogenization ... 64

4. Precondition Analysis and LP ... 67

4.1 Introduction ... 67
 4.1.1 Precondition Analysis ... 67
  4.1.1.1 LP ... 68
 4.1.2 The Organization of this Chapter ... 68
4.2 An Outline of LP ... 69
 4.2.1 Learning from Worked Examples - An Outline ... 69
  4.2.1.1 Worked Examples ... 69
  4.2.1.2 Learning from Worked Examples ... 69
 4.2.2 Solving Equations - An Outline ... 72
4.3 Basic Structures of LP ... 73
 4.3.1 The Arrangement of Worked Examples ... 74
 4.3.2 LP Methods ... 76
  4.3.2.1 Preconditions and STRIPS ... 76

4.3.2.2 LP's Initial Methods	77
4.3.2.3 The Key Methods	78
4.3.3 Rewrite Rules	79
4.3.4 The Condition and Connection Tables	80
4.4 How LP Learns	80
4.4.1 Operator Identification	81
4.4.1.1 Conjecturing an Instance of the Rule	82
4.4.1.2 Acquiring new rewrite rules	83
4.4.1.3 Storing the New Rule	84
4.4.1.4 Proof Checking	84
4.4.2 The Precondition Analysis Phase	85
4.4.2.1 Creating New Methods	88
4.4.2.2 Updating the Table of Connections	89
4.4.2.3 Choices in Applying a New Method	89
4.4.2.4 Rule Lists	90
4.4.2.5 When the Set ME is empty	91
4.4.2.6 The Division of the Worked Example	92
4.4.2.7 Schema Formation	93
4.4.2.8 Storing the Schema	93
4.4.2.9 The Strategy of the Schema	95
4.4.3 Plan Recognition	96
4.5 Solving New Tasks	97
4.5.1 Choosing the Appropriate Schema	97
4.5.1.1 Choosing a Subschema	98
4.5.2 Following The Schema	98
4.5.2.1 Is the Equation Solved?	99
4.5.2.2 Achieving the Purpose of a Section	100
4.5.2.3 Applying the Current Indicated Method	101
4.5.2.4 Substituting Operators	101
4.5.2.5 Adding Steps	103
4.5.2.6 Omitting Steps	104
4.5.2.7 Starting Again	104
4.5.3 Solving Problems without the Schema	104
4.5.3.1 Why Return to the Given Equation?	105
4.5.3.2 Finding a Schema	105
4.5.3.3 Differences due to the Lack of a Schema	106
4.5.3.4 Reporting Success or Failure	106
4.5.4 Examples of Schema Modification	107
4.5.4.1 Adding Steps	107
4.5.4.2 Omitting Steps	108
4.5.4.3 Substituting Steps	108
4.5.4.4 Achieving the Purpose of a Section	109
4.5.5 Using the Solution Trace as a Worked Example	110
4.6 Learning Existing Methods	111
4.6.1 The Successes	112
4.6.1.1 Learning Collection	112
4.6.1.2 Learning Attraction	113
4.6.1.3 Learning Factorization Preparation	113
4.6.1.4 The Logarithmic Method	114
4.6.2 Learning the Key Methods	114
4.6.3 Less Successful: Learning Homogenization	114
4.6.4 Conclusions	114
4.7 Possible Extensions	115
4.7.1 Adding Concept Learning	115

4.7.1.1 Concept Learning Programs	115
4.7.1.2 Precondition Analysis and Empirical Concept Learning	115
4.7.2 Extending the Use of Schemas	117
4.7.2.1 Summary	118
4.8 The Applicability of Precondition Analysis	118
4.9 Evaluation and Results	120
4.9.1 Evaluation	120
4.9.1.1 The Problem	120
4.9.2 The Terms of Evaluation	121
4.9.3 Tests and Results	121
4.9.3.1 LP as an Expert	121
4.9.3.2 LP as a Novice	122
4.9.4 Conclusions	123
4.9.5 Learning the Initial Methods - Examples	123
4.9.5.1 Learning the Attraction Method	123
4.9.5.2 Learning New Isclation Rules	123
4.9.5.3 Preparing for Change Of Unknown	124
4.9.6 Trigonometric Factorization Examples	124
4.9.7 Using Hints	124
4.10 Conclusions	126
4.10.1 Precondition Analysis and Meta-level Inference	126
5. Related Work	**127**
5.1 The Use of Meta-Knowledge	127
5.1.1 GOLUX	128
5.1.2 FOL	128
5.1.3 TEIRESIAS	129
5.1.4 Lenat's Approach	129
5.1.5 Summary	131
5.2 ALEX	131
5.2.1 How ALEX Works	132
5.2.1.1 Example	132
5.2.1.2 Learn	132
5.2.1.3 Perform	134
5.2.2 Discussion	134
5.3 The STRIPS System	135
5.3.1 The Operators	135
5.3.2 Storing the Plan - Triangle Tables	136
5.3.3 Executing the Plan	138
5.3.4 Developments	139
5.3.5 Relation of STRIPS to LP	139
5.4 LEX2	140
5.4.1 The Empirical-Analytic Spectrum	140
5.4.2 LEX	141
5.4.2.1 Problems with LEX	141
5.4.3 Description of LEX2	142
5.4.4 An Example	143
5.4.4.1 Producing the Explanation	144
5.4.4.2 Extracting Sufficient Conditions	145
5.4.4.3 Restating the Condition	145
5.4.4.4 Propagating the Restrictions	146
5.4.4.5 Learning New Concepts, Utgoff's Work	148
5.4.5 Comparison with Precondition Analysis	149
5.4.5.1 Sequence Learning	149

	5.4.5.2 Applicability of Mitchell's Method	149
	5.4.6 LEX2 Conclusions	150
	5.4.6.1 Other Work	151
5.5	Summary of Chapter	151

6. Summary   153

6.1	Constraining Search	153
6.2	Learning	154
	6.2.1 Other Learning Programs	155
6.3	Further Work	155
	6.3.1 New Domains	155
	6.3.1.1 Precondition Analysis	156
	6.3.2 The Self-Improving Algebra System	156
	6.3.3 Analytic Concept Learning for LP	156
6.4	Conclusions	157

Appendix I. The Specialist Methods   159

I.1	Introduction	159
I.2	The Nasty Function Eliminators	160
	I.2.1 Function Stripping	160
	I.2.2 Quotient Removal	161
	I.2.3 Eliminating Radicals	163
	I.2.3.1 Function-Isolation	163
	I.2.3.2 Inverse-Cancellation	164
	I.2.3.3 Argument-Reduction	165
	I.2.4 Inverse-Trigonometric Removal	165
	I.2.4.1 Function-Swapping	166
	I.2.4.2 A More Complex Example	166
	I.2.4.3 Function-Collection	167
	I.2.5 Summary of the Nasty Functions	167
I.3	Trigonometric Factorization	167
	I.3.1 The Two Term Case	169
	I.3.1.1 The Tangent Method	169
	I.3.1.2 Equations with Only One Type of Function	170
	I.3.1.3 Equations with equal angles	170
	I.3.1.4 Equations with additive angles	171
	I.3.2 The Mixed Case	171
I.4	The Logarithmic Method	172
I.5	Summary	173
I.6	Proving that an Expression is an Identity	173

Appendix II. Homogenization Specialists   177

II.1	Exponential Offenders Sets	177
	II.1.1 The First Exponential Case	178
	II.1.2 The Second Exponential Case	178
II.2	The Logarithmic Case	179
	II.2.1 The First Logarithmic Case	179
	II.2.2 The Second Logarithmic Case	180
	II.2.3 The Third Logarithmic Case	180
	II.2.3.1 The Power Method	180
	II.2.3.2 The Base 10 Method	181
	II.2.4 The Fourth Logarithmic Case	181
II.3	The Hyperbolic Offenders Set	181
II.4	The Exponential and Hyperbolic Offenders Set	182

Appendix III. Generating New Problems   183

Appendix IV. PRESS Output — 185
Appendix V. LP Output — 193

LIST OF FIGURES

Figure 1-1:	Some typical A level equations	8
Figure 2-1:	Expression tree for equation (vi)	17
Figure 2-2:	Least covering tree of the unknowns	18
Figure 2-3:	Worked Example for equation (vi)	19
Figure 2-4:	More Equations That Can Be Solved By PRESS	32
Figure 3-1:	Some Equations that are solved by Homogenization	50
Figure 4-1:	A Worked Example	70
Figure 4-2:	A Worked Example (repeated)	74
Figure 4-3:	The Worked Example after Operator Identification	85
Figure 4-4:	The Division of the Worked Example	94
Figure 4-5:	Part of the Schema	95
Figure 4-6:	Flow Chart for Following Schema	100
Figure 4-7:	A New Worked Example	110
Figure 4-8:	Part of the Schema	111
Figure 5-1:	A triangle table	136
Figure 5-2:	A Worked Example	143
Figure 5-3:	Explanation Tree for Example 5-2	144
Figure I-1:	Radical Removal Example	163
Figure I-2:	Trigonometric Factorization Equations	168

LIST OF TABLES

Table 2-1:	Characteristics of the Methods Described in this Chapter	33
Table 4-1:	Analysis of Worked Example	88
Table 4-2:	Results of Evaluation	122
Table I-1:	Characteristics of the Nasty Function Methods	168
Table I-2:	Characteristics of the Methods Described in this Appendix	174

CHAPTER 1

INTRODUCTION AND OVERVIEW

1.1 The Problems

This book addresses two questions:

- How can search be controlled in domains with a large search space?

- How can this control information be learned?

It is argued that both problems can be tackled with the aid of a technique called **meta-level inference**. Two programs are presented in this book to support this claim.[1]

The first program, PRESS[2], described in chapter 2, solves symbolic, transcendental, non-differential equations. The second program, LP[3], described in chapter 4 is also an equation solving program, but, unlike PRESS, it is capable of learning new equation-solving techniques. It embodies a new learning method, called **Precondition Analysis**. Precondition Analysis combines meta-level inference with concepts from the field of planning.

Why choose equation solving as the domain? The short answer is that equation solving has a very large search space, yet humans can perform well in this domain. The large search space is necessary to test the utility of meta-level inference, it also implies that the automation of equation solving is difficult enough to be interesting. The fact that we can solve equations means that we may be able to provide enough insight to make the automation problem tractable. Section 1.4 provides more discussion on the choice of domain.

1.1.1 Search

Most programs in Artificial Intelligence (A.I.) have to contend with the same problem, the search space is too large to be searched exhaustively. Therefore, these programs can search only part of the space. The difficulty lies in deciding **which** part should be searched, the program needs some way of constraining the search process.

---

[1]*Authorship*: LP has been written entirely by the author. The PRESS program has been modified over several years by a team including the author. PRESS and LP are written in DEC-10 Prolog, [23, 82, 5].

[2]$\underline{P}$rolog $\underline{E}$quation $\underline{S}$olving $\underline{S}$ystem.

[3]$\underline{L}$earning $\underline{P}$RESS.

Many simple approaches are possible. For example, using any search strategy,[4] a depth bound can be set that limits how deeply the program can search the space. The trouble with this kind of approach is that it is very inflexible. A uniform depth bound causes difficulty in that different problems need different depth bounds. One way of overcoming this is to give a new depth bound for every problem. However, such a solution is very ad-hoc.

What is needed is a method of constraining search that is sufficiently flexible to take account of the current state of the problem. Meta-level inference appears to provide such a solution.

1.1.2 Learning

Workers in Artificial Intelligence have long been interested in learning. At first, there were attempts to build general-purpose learning systems, i.e. programs that can learn in any domain. However, these programs did not really learn very well. One of the major difficulties was that the workers did not know *what* these programs should learn, as they lacked experience with problem-solving systems. McCarthy, [50], summed up the difficulty:

> "In order for a program to be capable of learning something, it must first be capable of being told it."

As the early systems lacked this knowledge, they did not improve in performance. These programs were not learning the right things. Usually they were learning far too little due to the over-optimism of the authors. Additionally, the authors had no particular domain in mind, they were attempting to learn general problem-solving techniques.

The conclusion was that the problem of learning should be put aside, until workers in A.I. had a good understanding of techniques of problem solving.

Recently, there has been a great increase in the number of learning programs being reported in the literature. This reflects the feeling that A.I. has built up a good grasp of what problem-solving involves.[5] This knowledge enables us to build powerful problem solvers, and also tells us what needs to be learned. Today, most successful learning programs include a "problem-solving element" and a "learning element". The problem-solving element attempts a task in some domain, and the learning element modifies the behaviour of the problem-solver in some way, so that the program is able to solve more problems or to solve them better in some way.

Another important reason for the new interest is that new problem-solving techniques sometimes suggest new learning techniques. This has happened with LP: the success of a meta-level inference solution to search in problem-solving suggested the technique of Precondition Analysis to learn control information.

1.2 Meta-level Inference

What is **Meta-level Inference**? In this technique, the control information is separated from the factual information. The control information is expressed **declaratively**, i.e. the control information is represented as explicit rules. These rules are axioms in the meta-theory of the domain. This gives

---

[4] For a description of search strategies, see, for example, chapter 2 of [60].

[5] Some people have been examining the problem of "learning to learn", examining how a program can modify its learning techniques. In an echo of McCarthy, it has been suggested that this area should not be attempted until A.I. has a better grasp of learning.

rise to a two level program, the factual information forms the object-level and the control information forms the **meta-level**. Inference is performed at the meta-level, and this induces inference at the object-level. Search at the object-level is replaced by search at the meta-level.

In PRESS, the object-level consists of algebraic rewrite rules (directed identities) such as
$X^2 - 1 \rightarrow (X + 1) \cdot (X - 1)$.
The meta-level of PRESS consists of **methods**. Methods are operators that apply the rewrite rules. Each method contains

1. A set of preconditions which must be satisfied before the method can be applied.

2. An associated set of rewrite rules, all the rules in a set achieve the same syntactic subgoal. For example, every rule in the **Isolation** method set will reduce the depth of nesting of functions surrounding the equation. (Isolation is described in section 2.3.1.1.)

3. A procedure for the selective application of the rewrite rules.

For example, consider the following meta-level axiom from PRESS.
singleocc(X,L=R) & position(X,L,P) & isolate(P,L=R,Ans) $\rightarrow$
    solve(L=R,X,Ans).
This can be considered as a procedure, to solve the goal on the right of the arrow, satisfy the subgoals on the left. However, it also has a **declarative** meaning in the meta-theory. The declarative meaning is

"If $L = R$ contains exactly one occurrence of X, and the position of this occurrence in L is P, and if the result of isolating X in $L = R$ is Ans, then Ans is a solution to the equation $L = R$, with X as the unknown."

Note that the description refers to properties such as position and the number of occurrences of X. These are meta-theory syntactic features. The description can be viewed as the statement of the correctness of the Isolation method, if Isolation can be applied, then the result produced is a solution to the original equation.

The inference of PRESS occurs at the meta-level. Some of the meta-level predicates are of the form
    New is the result of applying Rule to Old.
To satisfy such a predicate, the rule Rule is applied to the expression Old to produce New. As a result of this, an algebraic transformation has occurred at the object-level, the expression Old has been transformed into New. Thus, in PRESS, search at the object-level is replaced by search at the meta-level. Bundy and Welham, in [21], argue that the technique of meta-level inference has four advantages:

1. The separation of the factual and control information enhances the clarity of the program and makes it more modular.

2. The search process can be controlled in a flexible way. Inference can be used to adapt to circumstances not explicitly foreseen by the programmer, the process is data-sensitive.

3. The meta-level search space is often smaller than that of the object level, and this reduces the problems of combinatorial explosions.

4. The modularity of the program enables the learning of both new factual knowledge and strategic knowledge. In particular, the formalisation of the control information makes clearer what is to be learnt.

The programs described in this book, PRESS and LP, support these claims, particularly points 2 and 4.

PRESS makes extensive use of meta-level inference, and demonstrates points 1, 2 and 3 above. Sections 2.2 and 2.5.1 discuss meta-level inference in PRESS.

LP illustrates all four of the above claims. In particular, LP's Precondition Analysis provides strong evidence for claim 4. The work with LP has demonstrated that some of the concepts of the planning field can be used to advantage in learning programs, and this is only possible because the control information is explicitly represented.

1.3 Precondition Analysis

A program using **Precondition Analysis** works in two phases, the learning cycle and the performance phase. In the learning cycle, the program is given an example of a correctly executed task. In the context of LP, this is a worked example showing a way of solving an equation.

LP learns three kinds of things in the learning cycle:

1. **Rewrite Rules**: These are algebraic facts such as
   $$\cos(x) + \cos(y) \implies 2 \cdot \cos((x+y)/2) \cdot \cos((x-y)/2)$$

2. **Methods**: As in PRESS, methods are the meta-level problem-solving operators which apply the rewrite rules. LP methods contain more control information than the PRESS methods somewhat more complex than the PRESS methods to enable planning, see section 4.3.2.1.

3. **Schemas**: These are equation solving plans, used by the performance element. A schema contains a list of the methods used in the worked example, annotated with the reasons found for applying each method. Recording the reasons allows LP to adapt the plan suitably during the equation solving process. Schemas are described in section 4.4.2.7.

Precondition Analysis relies on having control information explicitly represented, programs using meta-level inference have this property.

Some steps in a plan can be much more important than others, i.e. some steps *must* succeed for the plan to work, while others are merely suggestions. Precondition Analysis uses these distinctions, certain steps are labelled **major steps**, and portions of the schema are devoted to the task of making these steps succeed. By telling the program which are the really key steps, the program can patch plans that are failing, and can take advantage of fortuitous events.[6]

1.3.1 The Operators of LP

Many learning programs work in domains where the basic operators are similar to those of STRIPS, [35, 36]. LEX, [55, 56], SAGE, [44], and ALEX, [58] are examples. The operators of STRIPS have preconditions, an **add list** which contains the facts that the operator makes true, and the **delete list** which contains the facts that are no longer true after the application of the operator. The facts on the add list are sometimes called the **effects** of the method, the method makes these facts true. If the

---

[6] Hierarchical planners, such as ABSTRIPS, [63], also draw distinctions between the importance of various steps, see section 5.3.4.

preconditions are satisfied the operator can be applied, producing the effects.

The equivalent of operators in LP (and PRESS) are **methods**. However, methods do not have one of the desirable properties of STRIPS-type operators. The preconditions of a method represent necessary, but not sufficient conditions, whereas the preconditions of STRIPS-type operators are both sufficient and necessary.[7] In general, a method is not certain to succeed, even if the preconditions are applicable. This is because the preconditions are too general, but we can not give stronger preconditions that do not involve actually applying the method to test if it is applicable! Similarly, the effects of a method are hard to classify, it is difficult to specify the add and delete lists. It seems that this might be a problem in many domains.

Consider, for example, the task of catching a certain train. Suppose we want to find the preconditions etc. of the operator get_train(Person,Train), where Person and Train are variables. Certainly, the add list must contain a fact such as on_train(Person,Train) and a precondition should probably be that on_train(Person,Train) is not true. Can we find a set of preconditions that are both sufficient and necessary, assuming a fairly normal world? We start with obvious necessary conditions, e.g. at_station(Station,Person,Time,Train), where the train Train leaves station Station at time Time. The sufficient part is much harder. We have to consider a huge number of possibilities, e.g. not_cancelled(Train), not_diverted(Train), not_full(Train) etc. The temptation is to put all these conditions into one, such as catchable(Train). This really gains nothing, as all these conditions must be tested to ascertain the truth of catchable. This may take a lot of time to do, e.g. telephone the station to make sure the train is not cancelled, and checking that it is not diverted (yet), and some of them involve most of the work of applying the operator anyway (e.g. checking that the train is not full.) In such cases, the best action is to check that the necessary preconditions are satisfied, and then try to apply the operator. If the operator is successfully applied, then it can be applied! To see if you can catch the train, go to the station and try to catch it.

In summary, in some domains it seems better to give up the requirement that the preconditions be both necessary and sufficient, and find a set of conditions that are just necessary. To see if the operator is applicable, check the necessary preconditions and then just try it! To avoid wasting effort, the necessary preconditions should be as strong as possible. We do want to insist that we are at the station before we attempt to catch the train.

Similar remarks apply to the effects of the methods. LP uses **postconditions** rather than effects. The postconditions are used to specify what must be true after a method has been applied in the **desired** way, but there is no guarantee that the method will make these true, i.e. it may be possible to apply the method in an undesirable way, where the postconditions are false. Postconditions can be used to check that the method has been applied in the right way.

This fact that LP methods lack STRIPS-type properties is important. If the methods were as well-behaved as STRIPS operators, it would be possible to build a much more powerful planning system, and a lot of the heuristic nature of LP would be unnecessary.

There is also another reason why LP cannot use more traditional planning techniques, LP learns new operators. LP cannot be given all the detailed information about these new operators that many planning systems require. For example, LP is not given the preconditions of the new operators, it has to generate approximations itself. Precondition Analysis can be used in situations where little is known about the operators.

---

[7]Or rather, this is the usual state of affairs. Although STRIPS has some limited ability to cope with situations where the operator does not produce the intended effect, or cannot be applied, these are considered to be exceptional cases. See section 5.3 for further details.

Precondition Analysis is similar in many respects to other learning methods that are being investigated by other researchers, in particular Goal Directed Learning of Mitchell, [54], and the Explanatory Schema Acquisition of DeJong, [30, 31]. There are also several differences between these studies and the author's, see chapter 5 for more details.

1.4 The Choice of Domain

There is nothing new in using computers to solve equations. Computers have been used for this purpose for many years, indeed the first modern computers were designed for equation solving. Today, many scientists and engineers use computers solely for equation solving.

Why are workers in Artificial Intelligence interested in equation solving? Firstly, it should be noted that we are dealing with *symbolic* equations, the equations that concern the scientist and engineer are usually *numerical*.

The equations used by PRESS are questions from **A level** Mathematics papers. A levels are exams taken in England and Wales by 18 year olds. They are used as a criterion for admission to university. The equations are **R Elementary Equations**. These are transcendental expressions, i.e. those involving polynomial, exponential, trigonometric, logarithmic and hyperbolic functions. A formal definition can be found in [21]. Figure 1-1 below shows some examples of A level equations.

The difference between symbolic and numerical equations lies in the form of the answer required. For numerical equations, the answers must be numbers, or a set of numbers, e.g. $x = 3.12, y = 1.4$. The answer to a numerical equation will usually be only an approximation, accurate within some given tolerance.

The answer to a symbolic equation need not be a number. It can be, e.g. $x = \tan(2 \cdot a + 1)$. Methods for solving symbolic equations are usually associated with the domain of algebra, the solution of numerical equations forms part of the field of numerical analysis.

It should also be noted that the problem of solving a symbolic (R Elementary) equation in one unknown is recursively undecidable, see [10] for references. This means that no decision procedure exists that can correctly answer "yes" or "no" to the question

Is the expression Ans a solution to the equation Eqn?

when given Eqn and Ans as inputs. This also means, of course, that PRESS will not be able to solve all such equations.

In contrast, algorithms exist for many forms of numerical equations that will calculate answers to any desired accuracy.

This book is concerned only with symbolic equations. When discussing symbolic equations the word symbolic will usually be omitted.

1.4.1 Equation Solving and Search

Why is symbolic equation solving an interesting task for Artificial Intelligence? The task of solving an equation consists of taking an equation and transforming it using legal algebraic operations until a "goal" state is obtained. The equation is in the goal state when it is of the form[8] $x =$ ans, where x is

---

[8] We say that the equation has answer ans.

the unknown and ans does not contain x, or if it is a disjunction of such terms.

Equation solving is a difficult task because at each step, the number of possible legal operations is very large. In [9][9], Bundy shows that 10 is a conservative figure for the branching rate in this domain. Possible operations include the use of commutativity, identity, and functional reflexive axioms. (See section 2.5.2 for more details.) To obtain an estimate of the typical search space, we need to know the average length of the solution path. For the problems in the test set used by PRESS (see next section) 10 steps seems to be a reasonable estimate. This implies that the average search space contains ten thousand million nodes at the depth that the solution occurs. Of course, some of the paths rejoin, so the actual search space will not contain this number of different nodes. However, the number will still be very large.

A program that had to examine all these possibilities would soon get bogged down in what is known as the **combinatorial explosion**. As the number of steps in the solution increase, the number of possible paths increases exponentially. What is needed is some way of pruning the search space. The program should only consider a small subset of possible operators at various points in the search.[10]

Some of the possible operations are never useful and can be disregarded, for example multiplying both sides of the equation by 0.

Most of the other operations are useful in some circumstances, but useless or harmful in others, and can not be so easily pruned from the search space. However, despite the very large search space, skilled humans can find the correct solution path, with almost no search at all! Apparently, these people are using some technique of search constraint. One of the aims of the PRESS project was to try and implement a program with similar abilities, so that it too could solve equations.

Humans, of course, can not only solve equations, they can also learn new techniques for equation-solving. LP is able to do this to some extent.

1.4.2 A level questions

The early versions of PRESS demonstrated that meta-level inference successfully constrained search in the equation solving domains. These versions were tested on relatively simple equations, usually invented by the researchers to illustrate various techniques.

Since then, there has been interest in extending PRESS so that it can solve equations of greater difficulty. It was decided to use A level Mathematics papers as the source of equations.

A level equations were chosen as the standard for two reasons:

- The equations seem to be of the right level of difficulty. No existing program was capable of solving even a small fraction of the exam equations. While most people find the equations too difficult, A level students do manage to solve most of the problems.

- The questions are set every year by external bodies, that have no knowledge about the program. This is preferable to creating our own problems, as otherwise we may

---

[9] pp15-16.

[10] It is also desirable for the program to detect if its current approach is not leading anywhere useful, and in this case to search elsewhere. This is hard to implement, except by using arbitrary bounds on the size of the expression, or the depth of the search. PRESS does not have such a mechanism.

unconsciously choose only equations that PRESS could solve, and we may also not be aware of the full range of possibilities.

The equations used come from the A level examination papers set by 3 boards: Associated Examination Board (A.E.B), London and Oxford. In addition O level papers from the Oxford Board and Scottish H level papers are used. O levels are exams taken at 16 in England and Wales, H level is taken at 17 in Scotland. For humans at least, A level questions are harder than H level questions, which in turn are harder than O level questions.

Figure 1-1 shows some typical equations used by the PRESS program. PRESS can solve all the equations in the figure, the column marked Time gives the solution time in milliseconds, with PRESS running on a Dec KL-10. The table also shows the source of the question. Here, and in the whole of this book, unless otherwise indicated, the question comes from an A level paper.

Equation	Time (ms)	Source
$\cos(x) + \cos(3 \cdot x) + \cos(5 \cdot x) = 0$	1351	A.E.B. 1976
$e^{(3 \cdot x)} - 2 \cdot e^{x} - 3 \cdot e^{(-x)} = 0$	1892	London 1977
$\log_2 x + \log_x 2 = 5$	903	London 1978
$x^4 - 6 \cdot x^3 - 7 \cdot x^2 + 36 \cdot x + 36 = 0$	1079	Oxford 1979
$2 \cdot \cosh(2 \cdot x) + \sinh(x) = 0$	2996	London 1980

Figure 1-1: Some typical A level equations

1.5 Objectives and Motivations

This section describes the motivations and objectives behind PRESS and LP.

1.5.1 The Objectives of the PRESS program

Originally, a major purpose of PRESS was as a vehicle for testing the utility of meta-level inference as a means of constraining search. Although the early version of the program dealt with only a fairly small class of equations, the techniques developed appeared to show that meta-level inference could indeed be used to constrain search. In order to test these principles further, it was decided to extend the range of PRESS.

PRESS contains a set of algebraic manipulations techniques, called **methods**. PRESS has been extended by added new methods. Many of the early methods of PRESS are general-purpose but weak. Methods that were added later tend to be powerful, but special-purpose.

PRESS reflects a historical process in Artificial Intelligence in this respect. At first, workers in Artificial Intelligence concentrated on building all-purpose programs that used a few general techniques, GPS, [33], being a typical example. These general techniques are necessary, but more recently it has been recognized that they are rarely powerful enough to obtain high levels of performance. Many current programs also supplement their general techniques with a set of specialists (experts) that deal with special classes of problems. Some domains may require large numbers of specialists, and these can be difficult to design and integrate. It is therefore advantageous to try to learn these specialist methods from the general-purpose methods, this allows the program to generate new methods when required automatically.

The control flow of the powerful methods tends to be quite complex, and these methods especially demonstrate the advantages of meta-level inference. Specialist methods are described in chapter 3 and appendix I.

Apart from the above objective, it was also hoped that the program would perform sufficiently well to rival non-experimental "performance" equation solvers such as REDUCE, [39] and MACSYMA, [49]. PRESS is favourably compared with MACSYMA in section 2.6.

Both objectives seem to have been met to a large extent.

PRESS performs well as an equation solving program. Its performance is measured on a test set of A level Mathematics questions. It currently solves 86% of the A level test set, a figure comparable to very good A level students.[11] The performance aspect is discussed further in section 2.5.3.1.

The meta-level inference aspects are described in detail in chapters 2 and 3.

1.5.2 The Objectives of LP

The motivation for LP came from two directions:

- A general interest in learning programs, and the belief that a meta-level inference solution to the search problem enables learning.

- A desire to create a self-improving algebra system. A program that could automatically learn new equation solving techniques would be a starting point towards this goal.

LP has succeeded in learning new equation solving methods and schemas for several examples in the A level problem test set.

LP can be broadly classified as a program that learns from examples. However, the learning technique used by LP, Precondition Analysis, also uses some of the principles of planning. There are interesting differences between the algebra domain and the domains usually used by workers interested in planning, e.g. the blocks world, see [77, 64] for example. These differences are discussed in chapters 4 and 5.

Concept Learning

In general, learning programs not only acquire new techniques, they also refine existing ones, using concept learning methods. This refinement often transforms weak, general-purpose methods into powerful specialist ones. LP does not do this, it concentrates on the creation of new techniques and strategies. Section 4.7.1 discusses how concept learning could be added to LP.

---

[11] PRESS was developed using this test set, so this figure *should* be fairly high. However, we recently acquired a totally new test set of 49 A level questions, and with no modification PRESS is able to solve 41 of these, giving a figure of 83%.

1.5.2.1 The Self-Improving Algebra System

LP was originally conceived as part of a **self-improving algebra system**. Such a system should be able to do the following:

(a) Learn new methods of solving equations.

(b) Propose interesting new conjectures.

(c) Prove the correctness of these conjectures.

(d) Assimilate newly proved theorems into the program as rewrite rules.

LP is intended to provide ability (a), and it also provides part of (d) as well. Other parts of the system have been built. These include the theorem prover IMPRESS, [16, 75], (for ability (c)), and an implementation of the focussing algorithm of Plotkin, Young and Linz, [89, 20], based on Winston's concept learning program, [88] (for (b)). Davy's SCOPE (unpublished) also provides part of (d). However, the current version of LP is not very compatible with these programs, although its integration remains a goal for the future.

LP was also motivated by our experience with PRESS. Many new methods have been added to PRESS over a period of eight years. The development and integration of new methods is difficult, and LP can also be seen as an attempt to automate the process of adding new methods.

Some of the equations found on exam papers can be solved using "tricks", i.e. using algebraic manipulations that are only suitable for a very small class of equations. If PRESS contained such special methods for every class of equation, it would become a bag of such special-purpose tricks. Part of the motivation behind LP is to prevent this happening - let the program learn new specialist methods itself from more general methods.

LP has demonstrated that meta-level inference can aid the learning process, thus meeting the first objective. LP has also been successful in automating the process of acquiring new equation solving techniques. A full self-improving algebra system does not exist as yet.

1.6 The Relevance of this work to Other Domains

This work is concerned with the domain of equation solving, and section 1.4 above indicates why this domain is considered to be interesting. However, the work described in this book is applicable to other domains.

Firstly, the technique of meta-level inference seems to be applicable to any domain where search is a problem, i.e. almost all domains. Apart from the domain of equation solving, our group at Edinburgh has applied meta-level inference to mechanics, [17, 18], to theorem proving, in the program IMPRESS, [16, 75], and statistics, [61].

However, it may seem that the new equation-solving methods are not of relevance to other domains. This is not the case. In particular, uses in other domains have been discovered for the principal new method, Homogenization, described in chapter 3. For example, it appears that Homogenization can be useful in theorem proving, see section 3.3.

The learning program LP, described in chapter 4, embodies a new learning technique. It seems that this technique should be applicable to domains other than equation solving. This will be discussed further in chapter 4.

1.7 The Structure of this book

Adapting McCarthy's dictum (see section 1.1.2) before discussing how a program can learn new equation solving techniques, we must first describe what needs to be learned. To do this, the next two chapters of this book centre on PRESS, and describe how meta-level inference enables PRESS to solve equations without getting bogged down in search.

Chapter 2 describes an extended version of PRESS. It is extended in that structures are described that are not in PRESS, but are learned by LP. This is done to avoid having to break up the description of the learning process in chapter 4.

Chapter 3 describes one method in detail. The method described is Homogenization, a powerful, rather general-purpose technique, written by the author. Two types of Homogenization are described. The first half describes Standard Homogenization, which applies to single equations in one unknown. (The description of some of the more detailed procedures is continued in appendix II). The second half discusses Extended Homogenization, which applies to simultaneous equations. This chapter also describes how Homogenization can be applied to domains other than equation solving.

Chapter 4 describes Precondition Analysis, a learning technique used by LP, the author's learning program.

Chapter 5 discusses work that is related to the programs described in this book.

Chapter 6 contains a summary and the conclusions.

Appendix I describes four other equation solving methods written by the author. These methods are described in an appendix as, unlike other methods, they are not used much in other parts of the book.

Appendix II continues the description of the various subprocedures of Homogenization.

Appendix III describes the problem generator of LP.

Appendix IV contains examples of the output of the PRESS program, and appendix V shows output from LP.

1.7.1 Notation Conventions

Throughout this book, we must distinguish between algebraic (object-level) constants, algebraic variables, and meta-variables.

All object-level entities begin with a lower case letter, e.g. x, a. Object-level constants start with a letter near the beginning of the alphabet, object-level variables start with one near the end. Thus a, b and c are object-level constants, and x, y and z are object-level variables. This convention is of course common in elementary mathematics.

Meta-variables begin with an upper case letter, e.g. A. Note that *both* object-level variables and object-level constants are meta-level *constants*. Thus our convention for distinguishing between meta-variables and object-level entities follows the Dec-10 Prolog convention for variables and constants, where variables begin with an upper-case letter.

CHAPTER 2

AUTOMATED EQUATION SOLVING

This chapter describes the techniques used by PRESS to solve equations. Many of the methods described in this chapter are also used by LP, described in chapter 4. This chapter also contains descriptions of methods that are used by LP, but are not in the current version of PRESS. This is done to avoid having to break up the description of the learning process in chapter 4. The overall equation solving strategies of the two programs are different. In this chapter, the strategy described is generally that of PRESS.

Both programs are written in DEC-10 Prolog, [23, 82, 5].

Work on PRESS began in 1975, when Bob Welham wrote the first implementation, [83]. PRESS has been further developed by Alan Bundy, Leon Sterling, Richard O'Keefe, Lawrence Byrd, and the author.

Purpose of PRESS

PRESS was designed with two objectives in mind. The first was as a vehicle for exploring methods of constraining search, the second was as an equation solving module for **MECHO**, [18]. MECHO is a program that solves mechanics problems stated in English. The equations produced by MECHO are usually simple, and PRESS can solve them fairly easily.

This book discusses only the first objective of PRESS.

2.1 An overview of PRESS

PRESS is capable of solving equations in one unknown, simultaneous equations, inequalities, and sets of inequalities. This book deals only with equations.

The concepts of PRESS will be introduced in relation to equations in a single unknown. Simultaneous equations are discussed in section 3.2.

2.1.1 The Solution of an Equation

The goal of solving an equation is to obtain an expression called the **solution to the equation**. Suppose that the equation is lhs = rhs, and x is the unknown. Essentially, a solution of the equation has two properties. Firstly, it is an expression, or a disjunction of expressions, of the form

$x = a$, (i)

where a does not contain x. This is a purely syntactic property. Of course there is an infinite number of such terms, most of these are not solutions to the equations.

The other property relates to the standard model. In our case the model consists of the real numbers. Descriptions of the properties of the real numbers and the functions occurring in R Elementary equations can be found in textbooks on real analysis.

Substituting a for x in the equation lhs = rhs produces a relation of the form
$$\text{lhs'} = \text{rhs'}. \tag{ii}$$
If (i) *is* a solution to the equation lhs = rhs, (ii) is true in the standard model.[1] This distinguishes solutions from non-solutions.

2.1.2 Solving Equations

If the original equation is not already solved, PRESS attempts to solve it by transforming it into another equation[2], iterating this process until a solution is obtained.

We describe this process as successively transforming the equation, rather than replacing one equation with another. This allows us to talk about the current state of *the* equation.

An obvious requirement is that all the transformations performed use legal rules of algebra. Otherwise, substituting the solution into the equation may produce a statement that is false in the standard model. These legal moves are expressed as **rewrite rules**. Rewrite rules are directed algebraic identities, perhaps with conditions. For example, the identity
$$A = B \tag{iii}$$
can be transformed into the (unconditional) rewrite rule
$$A \Longrightarrow B, \tag{iv}$$
which means A can be rewritten to B.

2.2 Meta-level Inference in Equation Solving

2.2.1 The Exhaustive Application of Rewrite Rules

One possible approach towards equation solving is to have a set of rewrite rules, and to apply them exhaustively, i.e. to transform the equation using any rule that is applicable, and continue this process until no more rules apply. However, as the number of possible rules is so large, this technique can be quite useless. For example, it is legal in algebra to add the same number to both sides of the equation. This is sometimes useful, e.g. in completing the square. The equation
$$x^2 - 2 \cdot x = 3$$
can be transformed into
$$x^2 - 2 \cdot x + 1 = 3 + 1,$$
by adding 1 to both sides. This transformation uses the rewrite rule

---

[1] Of course, this is undecidable. There is no decision procedure that determines if the statement lhs = rhs is true. However, this is not usually a problem with A level questions!

[2] Sometimes the equation may be transformed into a conjunction or disjunction of equations.

$$A = B \Longrightarrow A + C = B + C,$$

with C instantiated to 1. Further operations simplify this to
$$(x - 1)^2 = 4,$$
leading to the solution
x = 3 or x = -1.

However, it is clearly undesirable to have a program that attempts to add 1 to both sides of any equation that it is given. Even if it did not do this repeatedly (an infinite loop), the step would nearly always be a waste of time.

Another problem with exhaustive rewriting is that algebraic identities can not be used both ways round. The identity (iii) can produce not only (iv), but also:
$$B \Longrightarrow A. \tag{v}$$

However, if both (iv) and (v) are present in the system, an exhaustive rewriting technique would loop, i.e. A is rewritten to B using (iv), then B is rewritten to A using (v), this process can continue indefinitely. To prevent this, exhaustive rewriting systems include only one of the rules (iv) and (v). However, if both rules are not included, the system may be incomplete, i.e. there will be equations that can not be solved by the system with its set of rules.

Mechanisms have been suggested for completing incomplete systems, see [41] and [40], but these techniques are not certain to terminate, so it is not possible to guarantee that a complete set can be formed. These techniques are not algorithmic in another way, the proofs of termination contain creative steps which are not mechanical. Often these creative steps involve finding a metric that decreases.

In summary, exhaustive rewriting systems have rules only one way round, and are incomplete. This obviously restricts the usefulness of such systems.

Bundy and Welham, [21], give some further problems encountered with exhaustive rewriting systems:

- Sets of rewrite rules often have an ad hoc character as there is no clear basis for including or excluding particular rules.

- A rule may be included because it is essential in one situation. In another situation the rule becomes an unwanted embarrassment.

As there are many problems caused by applying rewrite rules exhaustively, a different approach is needed. PRESS uses selectively applied partitioned rewrite rule sets as one element in the technique of meta-level inference.

2.2.2 The Selective Application of Rewrite Rules

There are two levels in a system that uses meta-level inference. The first, the object-level encodes the factual knowledge of the domain. The other level, the meta-level, encodes the control or strategic knowledge.

The object-level consists of rewrite rules. The rewrite rules are divided into several sets. Each set performs a particular kind of algebraic manipulation. Associated with each set is a syntactic characterization of the kind of rule it contains. This syntactic characterization is used by the inference process to determine which set of rules is suitable at a particular point in the solution process.

The syntactic characterizations provide a clear basis for deciding whether a rule should be included in a set or not. Thus the sets of rewrite rules of PRESS do not have an ad hoc flavour.

As there are multiple sets of rules, a particular rule may appear in each direction in different sets. Since the rules are selectively applied, the rule may even appear in both directions in the same set. In this case, the rule does not cause looping, as the rule is applied to different subterms in each direction, see section 2.3.1.3 for an example of this.

The problem of embarrassing rules disappears as well, as different sets are used in different situations.

The meta-level knowledge consists of a set of algebraic manipulation methods. Meta-level inference is used to choose the appropriate method, and the method uses meta-level inference to select and apply rewrite rules to the equation, i.e. the meta-level knowledge allows the rewrite rules to be applied selectively. Further discussion on the meta-level inference of PRESS requires a greater knowledge of the program. The following sections provide this detail. The discussion on the meta-level inference is resumed in section 2.5.1.

The next section describes the methods used by PRESS.

2.3 The Equation Solving Methods

PRESS has a set of algebraic manipulation techniques, called **methods**. The methods form the meta-level. Each method has:

1. A set of preconditions which must be satisfied before the method can be applied.[3]

2. An associated set of rewrite rules, all the rules in a set achieve the same syntactic subgoal.

3. A procedure for the selective application of the rewrite rules.

Some LP methods also contain schema information, used for planning. See chapter 4.

The preconditions and postconditions are meta-level goals, i.e. they test various syntactic features of the equation. The preconditions can also be viewed as **recognizers**, which test if the equation is in the domain of applicability of the method. See section 2.5.2.2 for more details.

Section 2.3.3 describes how PRESS uses the methods to solve equations. A similar technique is used by LP before learning takes place. Chapter 4 describes how learning modifies the behaviour of LP.

The individual methods will be described shortly, but the syntactic features must be discussed first. This section also defines some terms.

---

[3] In addition, LP methods have explicit postconditions, conditions which must be true **after** the application of the method. Postconditions differ from what are usually called **effects**, conditions made true by the method. The postconditions are used in learning, see chapter 4. As PRESS does not learn it does not need postconditions. Section 4.3.2.1 in chapter 4 contains more discussion of postconditions.

The Expression Tree

An equation is normally presented as a linear string of symbols. It is often more useful to view it as a two dimensional tree, the **expression tree**. At the root of the tree is the dominant symbol of the expression, normally the equals sign, $=$. The descendants of this node are the left and right hand sides of the equations. These nodes are similarly expanded. For example, figure 2-1 shows the expression tree for the equation

$$(2^{x^2})^{x^3} = e. \tag{vi}$$

(In the figure, "^" denotes the exponentiation operator.)

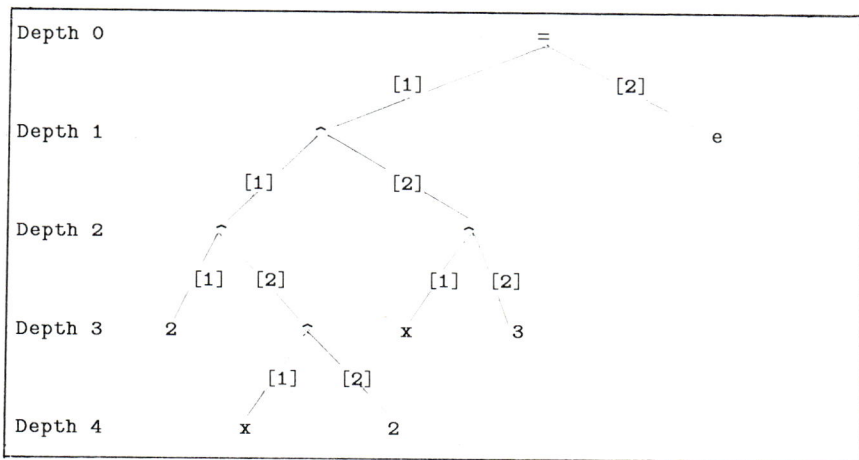

Figure 2-1: Expression tree for equation (vi)

The following terms can now be defined:

- The **depth** of a term: The depth of the term in the tree is defined in the usual recursive way, the root has depth 0, and a successor of a node of depth n has depth n + 1.

- The **size** of a subtree: The size of a subtree is simply the number of arcs contained in that subtree. (Note that all arcs have equal weight.)

- The **least covering tree** of a set of terms: This is the smallest subtree of the expression tree that includes each member of the set. The least covering tree of the two x terms in figure 2-1 is shown in figure 2-2.

The important concept of **position** is defined using depth. The **position** of a term is a list describing the path in the tree from the root to the term. The nth member of the list describes the path at depth n. The value of this entry is calculated by numbering the arcs radiating from a node from left to right, starting with the number 1.

As an example, we will calculate the position of the x in the x^2 in figure 2-1. The depth of each level is indicated on the left of the figure.

After the root, at depth one the node labelled ^ is chosen, rather than the node labelled **e**. This is the left most node and is reached by going down the arc labelled 1. At depth 2, the left-hand exponentiation term must be chosen, traversing arc number 1. At depth 3 the chosen node is the one labelled ^ rather than the **2** node. This lies on the arc labelled **2**. At depth 4, the node labelled **x** is chosen, which lies on arc 1. number 1. We have now arrived at the x in the x^2 term, so the desired position is $[1,1,2,1]$.

The least covering tree is used to define **closeness**, which is the size of the least covering tree of the unknowns. Consider the tree in 2-1. The least covering tree of the unknowns is shown in figure 2-2. This has five arcs so the closeness is 5.

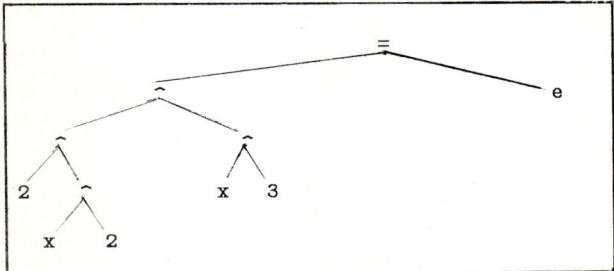

Figure 2-2: Least covering tree of the unknowns

The following concepts are also used extensively:

- The number of occurrences of the unknown in the equation.

- The positions of the unknowns in the equation.

- The type of function symbols occurring in the equation.

2.3.1 The Individual Methods

This section describes the individual methods. Except where otherwise indicated, both PRESS and LP contain the method described. PRESS has a fixed set of methods, LP can learn new ones.

Note: The first six methods described, (Isolation, Collection, Attraction, Factorization, Change of Unknown and Polynomial Methods), were not designed by the author.

The first example illustrates three methods, Isolation, Collection, Attraction. Figure 2-3 shows how to solve the equation (vi).

$(2^{x^2})^{x^3} = e$

This equation is a modified version of an example in [21].

2.3.1.1 Isolation

It is easier to examine the example by working backwards from the end. The last line of the example, line (x), is the solution to the equation. There is a single occurrence of x, occurring at position [1].

The previous line (ix) also has the only occurrence of x on the left hand side, but in this case it is at position [1,1], as it appears as the first argument of the exponentiation function. The last line is obtained from (ix) by applying the inverse of this function to both sides.

$$(2^{x^2})^{x^3} = e$$

$$(2^{x^2 \cdot x^3}) = e \qquad \text{(vii)}$$

$$2^{x^5} = e \qquad \text{(viii)}$$

$$x^5 = \log_2 e \qquad \text{(ix)}$$

$$x = (\log_2 e)^{(1/5)} \qquad \text{(x)}$$

Figure 2-3: Worked Example for equation (vi)

In the case of line (ix), we say that the exponentiation function **dominates** the occurrence of x, or that exponentiation is the **outermost** function. The outermost function of an occurrence of a term is the function referred to by the second member of the position list (the first refers to the equality predicate).

Similarly, line (ix) is obtained from line (viii) by removing the outermost function, which is also exponentiation in this case. (This time, the term containing x is the exponent rather than the base.) A function is removed by applying the inverse of that function to both sides of the equation. In this case this is done by taking logarithms to base 2.

This process of removing the outermost function is called **Isolation**. Isolation works on equations containing exactly one occurrence of the unknown. At each application it removes the outermost function surrounding the occurrence of the unknown, until the unknown appears alone.

Isolation rules are of the form

$$C(U_1, U_2, \ldots, U_n, B), \& \; F(U_1, U_2, \ldots, U_n) = B \rightarrow U_i = F_i^{-1}(B),$$

where C is a (possibly empty) set of conditions, U_i contains the unknown, and F_i^{-1} is the ith inverse of F.[4] For example, the Isolation step from (viii) to (ix) uses the rule

$$A^U = B \rightarrow U = \log_A B.$$

2.3.1.2 Collection

Now consider the step from line (vii) to line (viii). The number of occurrences of the unknown reduces from two to one. This step is an example of **Collection**. Collection can only be applied to equations containing two or more occurrences of the unknown, and its effect is to reduce the number of occurrences. Note that Collection need not reduce the number to one, so Isolation may not apply after Collection. Sometimes a sequence of Collection applications will be needed to reduce the number of occurrences to one, so that Isolation can be applied, in other cases totally different methods will be needed.

The Collection rule used in the example is

$$U^V \cdot U^W \Rightarrow U^{V \cdot W}$$

---

[4] In some cases there will be a disjunction of terms on the right-hand side. For example, if F is the square function, $x^2 = a \rightarrow x = a^{1/2} \; \vee \; x = -(a^{1/2})$.

Other Collection rules include:
$$U \cdot N + U \cdot M \Longrightarrow U \cdot (N + M)$$

$$U^2 + 2 \cdot U \cdot V + V^2 \Longrightarrow (U + V)^2$$

In the first case the term corresponding to U must contain the unknown. Either U or V can be used in the second case.

Note that Collection rules only need to be applied to particular subterms of the equation. Collection is applied only to a term that is **least-dominating** in the unknown. A least-dominating term in x is a term with at least two immediate subterms both of which contain x. Thus
$$(x + 1) \cdot (x - 1)$$
is least-dominating in x, whereas
$$\log_e((x + 1) \cdot (x - 1))$$
is not least-dominating, as only one of its immediate subterms contains x, the other subterm being e. Similarly, $x + 1$ is not least-dominating, as only one of the immediate subterms contains x.

A combinatorial explosion would result if rewrite rules could be applied to any subterm of the equation. This problem is prevented by restricting rewrites to least-dominating terms. Again, this is an example of the selective application of rewrite rules, using meta-level inference. Note that Isolation is similarly restricted, at any stage it strips off only the outermost function dominating the unknown. For example, given the equation
$$\sin(\cos(x)) = \tan(a),$$
Isolation strips off the sin function rather than the tan one.

There is no power lost by the restriction to least-dominating subterms. If a Collection-type rule applies to a term which is not least-dominating, a form of Isolation can be applied to the rule to produce a new rule that does apply to a least-dominating subterm.

For example, applying the rule
$$2 \cdot (\sin(X) \cdot \cos(X)) \Longrightarrow \sin(2 \cdot X) \qquad \text{(xi)}$$
reduces the number of occurrences of X. However, it applies to terms that are not least dominating. The rule is equivalent to the Collection rewrite rule
$$\sin(X) \cdot \cos(X) \Longrightarrow \sin(2 \cdot X)/2,$$
obtained by removing the outermost functions from both sides of (xi), a form of Isolation.

Firstly, PRESS tries to apply Collection to the least-dominating subterm of the whole equation. If this fails, the equation is broken up into subterms, and Collection is attempted on least-dominating subterms of these subterms.

For example, suppose the equation is
$$\sin(2 \cdot x) + \sin(x + x) = 1.$$
The least-dominating subterm of the whole equation is the LHS. PRESS does not have any Collection rules for this term, so Collection can not be applied. The LHS is then broken into the subterms
$$\sin(2 \cdot x) \text{ and } \sin(x + x).$$

The first of these subterms has no least-dominating term, as it only contains one occurrence of the unknown. The second subterm has the least-dominating subterm $x + x$. Collection can be applied, transforming this to $2 \cdot x$. This produces the equation
$$\sin(2 \cdot x) + \sin(2 \cdot x) = 1.$$

Now Collection can be applied to the whole LHS, producing an equation that is solved by Isolation.[5]

When the equation is large, there may be many least-dominating subterms that Collection has to consider. Collection can take a large amount of time on such equations. However, the search space is still very much smaller than if Collection had to consider *all* subterms, whether least-dominating or not.

2.3.1.3 Attraction

This leaves the step from the original equation to line (vii),

$(2^{x^2})^{x^3} = e$

$(2^{x^2 \cdot x^3}) = e.$

The number of occurrences of the unknown does not reduce, so it is not a Collection step. A calculation of the closeness of the equations shows that the size of the least covering tree *does* reduce. In the original equation the closeness is 5, in (vii) it is 4. Collection is more likely to be applicable if the terms are brought closer together.

This step is an example of **Attraction**. The Attraction rewrite rule used is:

$(U^V)^W \implies U^{(V \cdot W)}$ (xii)

which brings the V and W terms closer together.

Attraction can only apply to equations containing at least two occurrences of the unknown. Its effect is to reduce the closeness of the equation. Note that Collection may not be applicable after an Attraction step, the purpose of Attraction is just to make this more likely.

The remarks made about least-dominating terms in section 2.3.1.2 apply to Attraction rules as well, i.e. Attraction is applied only to terms that are least-dominating in the unknown.

Experience with PRESS has shown that Attraction is a weak method compared with Isolation and Collection. Attraction is rarely applicable, and its application does not always greatly increase the chance that the equation will be solved. However, there are occasions when Attraction is needed, as shown in the step above.

Other Attraction rules include:

$\log_W U + \log_W U \implies \log_W U \cdot V$

$U \cdot W + V \cdot W \implies (U + V) \cdot W$

$U^{(V \cdot W)} \implies (U^V)^W$ (xiii)

In these rules, Attraction is applied to the terms U and V, so these terms must contain the unknown, and the other terms must not.

---

[5] Alternatively, Change of Unknown can be applied, producing a polynomial.

Prevention of Looping

Note that rule (xiii) is the same as rule (xii) above, applied right to left. An exhaustive rewriting system could not have both rules, as their presence would cause looping. In PRESS, these rules are only applied for the purpose of Attraction, to reduce the closeness of the equation. Suppose that rule (xii) is applied to Old to obtain New. For this to be a valid Attraction step, the closeness of Old must be greater than that of New. Now, if rule (xiii) is applied, New is replaced by Old. This increases the closeness measure, so the step is not a valid application of Attraction. Therefore the rewrite is not performed, and no looping occurs.

2.3.1.4 The Role of the Worked Example

From figure 2-3, we have identified three methods, Isolation, Collection, and Attraction. The worked example contains an instance of each (two in the case of Isolation), and from the characteristics of these steps it is possible to construct a set of appropriate rules. In other words, the worked example enables us to define methods.

The worked example also demonstrates how these three methods fit together. The solution is produced by an Isolation step, if Isolation can be applied it will lead to a solution. However, it isn't always possible to apply Isolation, the equation can contain more than one occurrence of the unknown. If this is the case, Collection can be helpful in reducing the number of occurrences of the unknown. Similarly, if Collection cannot be applied at once, Attraction may help.

So from the worked example, it is possible to construct a general plan. In [21], Bundy and Welham call this plan the **Basic Method** of PRESS. However, PRESS does not have an explicit representation of this plan, it is implemented by the priority ordering of the methods, see section 2.3.3. In contrast, LP, described in chapter, 4, does explicitly represent plans using the technique of Precondition Analysis. This allows more complex plans to be constructed, and the plans can be executed in a more directed way.

The following sections describe some of the other methods of PRESS. These methods were not required for the simple example above.

2.3.1.5 Factorization

An equation of the form
$$a \cdot b \cdot c \cdot \ldots \cdot k = 0 \qquad \text{(xiv)}$$
is equivalent to the disjunctive set of equations
$$a = 0 \lor b = 0 \lor \ldots \lor k = 0. \qquad \text{(xv)}$$

Each member of the disjunct (xv) can be solved as an independent equation. The solution of the original equation (xiv) is the disjunction of the solutions to the individual equations in (xv).

PRESS is able to perform this type of transformation, using its **Factorization** method. If the equation is of the form (xiv), PRESS finds the members of the product, and removes the ones that are non-zero constants.[6] The remaining terms are set equal to zero, and a disjunct of the form (xv) is

---

[6]This is done for two reasons. Firstly, we aim to make the output of PRESS resemble that produced by humans (although we don't claim psychological validity, but see section 2.7). If the non-zero factors were left in, PRESS would have to solve equations such as $2 = 0$. Humans don't seriously attempt to solve such equations.

The other consideration is efficiency. If PRESS has to solve $2 = 0$, for example, it produces the answer **false**, which would then go into the disjunction. Later processing would remove this from the final answer. It is more efficient to remove the factor before the solution process begins.

produced. The individual equations are then solved, and the disjunction of the solutions is returned. For example, consider the equation

$$3 \cdot \sin(x) \cdot \cos(x) = 0. \tag{xvi}$$

The 3 is removed, and the disjunct

$$\sin(x) = 0 \text{ v } \cos(x) = 0$$

is formed. PRESS then attempts to solve the two equations. In this case Isolation produces the answer

$$x = 180 \cdot n_1$$

for the first equation, and

$$x = 180 \cdot n_2 + 90$$

for the second. Thus PRESS returns the answer

$$x = 180 \cdot n_1 \text{ v } x = 180 \cdot n_2 + 90$$

as the solution to the equation (xvi).

Note that Factorization cannot do more sophisticated factorization, such as factorizing polynomials containing non-numeric constants,[7] e.g. factorizing

$$x^2 - (a + b) \cdot x + a \cdot b = 0$$

into

$$(x - a) \cdot (x - b) = 0.$$

PRESS at present has no method for performing this kind of factorization, such a method may be added in the future.

2.3.1.6 Polynomial Solution Methods

PRESS is able to solve a wide variety of polynomial equations, using its **Polynomial Methods**. Polynomial equations are important because many equations can eventually be transformed into polynomial form.

The polynomial methods are intended to be self-contained. If an equation is a polynomial, the polynomial methods will try to solve it, and no other methods are tried. This leads to some apparent duplication (see footnotes below) but it is conceptually cleaner to have one set of methods dealing with a particular class of equations.

PRESS is able to solve linear, quadratic and symmetric equations. (Symmetric polynomial equations are nth degree polynomials, where the coefficient of x^r is the same in magnitude of that of x^{n-r}, for all r between 0 and n. Such equations can be rewritten as a product involving a factor of $x + 1/x$ or $x - 1/x$ or both.) PRESS attempts to transform other polynomial equations into one of these types.[8]

For example, PRESS can recognize that

$$x^4 - 2 \cdot x^2 + 1 = 0$$

---

[7] The Polynomial methods can factorize *numeric* polynomials, see below.

[8] This process is really a special case of Homogenization, (see chapter 3), but it is done by the polynomial methods here. See remarks above.

is really a quadratic in x^2. Similarly
$$x^3 + 4 \cdot x^2 + 4 \cdot x = 0$$
is discovered to be a product
$$x \cdot (x^2 + 4 \cdot x + 4) = 0$$
which is factored into a linear equation and a quadratic one.[9]

If all its other methods fail, and the polynomial contains only rational numeric coefficients, PRESS uses the remainder theorem to try to find a root. (This method was added by the author.) Consider a polynomial equation
$$a_0 \cdot x^n + a_1 \cdot x^{(n-1)} + \ldots + a_{(n-1)} \cdot x + a_n = 0,$$
where all the a_i are rationals. Let the greatest common divisor[10] of the a_i be g. A root of this equation must divide $(a_n)/g$. PRESS finds all the divisors of this number, and evaluates the polynomial at each of the divisors. If the value of the polynomial at $x = c$ is 0, the remainder theorem states that $(x - c)$ is a factor of the polynomial. If such a factor is found, PRESS divides it out, and attempts to solve the remaining factor.

Note that this method is only used if the coefficients are rational numbers. In theory, it could be adapted to work on equations such as
$$x^2 - (a + b) \cdot x + a \cdot b = 0, \qquad \text{(xvii)}$$
by considering the factors of $a \cdot b$. There are considerable problems however. Suppose that the program first attempts to see if a is a root. Substituting a for x in (xvii) produces
$$a^2 - (a + b) \cdot a + a \cdot b = 0.$$
To discover if the left-hand side does equal 0, the program has to use PRESS methods, such as Collection, treating the constant a as the unknown. If this was allowed in general, the program might suffer from a combinatorial explosion, PRESS methods currently apply rewrite rules *selectively*, e.g. only to least-dominating subterms. Also, of course, the problem of deciding if an arbitrary expression evaluates to 0 is undecidable.

2.3.1.7 Change of Unknown

A common method of solving equations is to change the unknown, e.g. if the equation is
$$4/\log_2 x + \log_2 x = 5$$
substituting $y = \log_2 x$ gives the equation $4/y + y = 5$.

This is a disguised quadratic, as the equation can be transformed to
$$y^2 - 5 \cdot y + 4 = 0.$$

This equation can be solved to obtain $y = 1$ or $y = 4$. Now the equations
$$\log_2 x = 1$$

and

---

[9] This is the polynomial case of Factorization Preparation, described below in section 2.3.1.8, but it is done by the polynomial methods here. See remarks above.

[10] To find the g.c.d. of rationals express the rationals in terms of the lowest common denominator, k say, and take the g.c.d. of the resulting numerators, h say. The g.c.d. of the rationals is then h/k, reduced to the lowest terms.

$\log_2 x = 4$
may be solved to give the solutions for x
$x = 2$ or $x = 16$.

The above equation solving process is referred to as the **Change of Unknown Method**. The steps in the method are as follows:

(i) The equation is examined to see if all the occurrences of the unknown x are in identical subterms. In the example above the subterm is $\log_2 x$. There are two extra conditions. The first requirement is that the subterm is different from the unknown. Secondly, the subterm must occur at least twice in the equation. The justification for these conditions is given below.

If these conditions are not satisfied exit with failure.

(ii) Let us call the subterm found above f(x). Substituting y for f(x) in the equation the gives a new equation, the **changed variable equation**. This equation is solved to give a value or, set of values, for y, $y = y_i$ say. Now the equation(s) $f(x) = y_i$ are solved. The equation(s) of this form is (are) called the **substitution equation(s)**.

We must now justify the two extra conditions in step (i).

Note that in the example all the occurrences of the unknown *are* in identical subterms, namely x. This is of course true for any equation. However, y can not usefully be substituted for x, as the resulting changed variable equation is a trivial renaming of the original. Hence the need for the condition that f(x) be different from x.

The second condition, that the subterm occurs at least twice in the equation, is needed to ensure that the *substitution* equation is different from the original equation. Suppose that in the example the substitution
$y = 4/\log_2 x + \log_2 x$
was used. The changed variable equation is $y = 5$, which is trivially solved to produce the answer $y = 5$. Now the substitution equation is
$4/\log_2 x + \log_2 x = 5$,
which is identical to the original equation.

Substitution and Changed Variable Equations

Different equations can give rise to the same changed variable equation. For example
$3 \cdot \cos^2(x) + 4 \cdot \cos(x) - 7 = 0$ and
$3 \cdot (e^x)^2 + 4 \cdot e^x - 7 = 0$
both have the same changed variable equation
$3 \cdot y^2 + 4 \cdot y - 7 = 0$.

Of course, the equations are not satisfied by the same values of x. This is reflected by the fact that the substitution equations are different.

So, providing that both the changed variable and substitution equation can be solved, the Change of Unknown Method gives the solution to the original equation. However, the Change of Unknown Method relies on the unknown occurring within identical subterms in the original equation.

Unfortunately, things are seldom as simple as this. Equations, which can be solved by Change of Unknown, are more likely to appear in the form

$$4 \cdot \log_x 2 + \log_2 x = 5 \qquad \text{(A Level London 1978)} \qquad \text{(xviii)}$$

in which the occurrences of the unknown, x, appear within dissimilar subterms, namely $\log_x 2$ and $\log_2 x$.

Some preparation of the equation is required before the unknown can be changed. In the case of the example, one possibility is to convert the subterm $\log_x 2$ to $1/\log_2 x$, with the aid of the rewrite rule $\log_U V \Longrightarrow 1/\log_V U$.
This produces the equation shown at the beginning of this section,
$$4/\log_2 x + \log_2 x = 5.$$

Thus Change of Unknown is directly applicable to only a small number of equations. However, the author's method Homogenization, described in chapter 3, attempts to transform the equation so that Change of Unknown can be applied. For example, Homogenization would perform the rewrite in the example above.

Change of Unknown produces two subgoals, the Changed Variable equation and the Substitution equation. Other methods, such as Factorization, may produce more than one subgoal, but Change of Unknown is different in that it produces two subgoals that are not independent. The Changed Variable and Substitution equations share a variable, the new unknown. However, this turns out to be unimportant, the Changed Variable equation is solved first, and the result substituted into the Substitution equation.

2.3.1.8 Factorization Preparation

As the name suggests, **Factorization Preparation** transforms the equation so that Factorization can be applied.

This method takes an equation of the form
$$e_1 + e_2 + \ldots + e_n = 0,$$
and transforms it to
$$f \cdot (e'_1 + e'_2 + \ldots + e'_n) = 0, \qquad \text{(xix)}$$
where f is a common factor of all the e_i, and
$$f \cdot e'_i = e_i.$$
the e_i and f must contain the unknown. f is called the **common subterm** of the left-hand side.

This method uses the distributive law
$$A \cdot B + A \cdot C + \ldots + A \cdot M \Longrightarrow A \cdot (B + C + \ldots + M). \qquad \text{(xx)}$$

Note that Factorization can now be applied to equation (xix), this is the point of Factorization Method.

The above description implies a set of preconditions that must be satisfied before Factorization Preparation can apply. The right-hand side of the equation must be 0, the left-hand side must be a sum of terms containing the unknown and all the terms in the sum must have a common subterm. Note that the common subterm must contain the unknown, and must be a (top-level) multiplicative factor of each member of the sum. This last condition implies that the distributive law can be applied, and prevents x being counted as a common factor of
$$\cos(x) + \cos(2 \cdot x) + \cos(3 \cdot x).$$

This method was designed for LP. The current implementation of PRESS does not have an explicit Factorization Preparation method, but this just reflects the fact that LP was written after PRESS. Future versions of PRESS will probably contain such a method. At present, PRESS handles this case by Collection, note that the rewrite rule (xx) is a Collection rule if A contains the unknown.

A more general method?

Other rules apart from (xx) can be used to allow Factorization, but these are not included as they don't satisfy all of the above preconditions. For example, the rule
$$\cos(A) + \cos(B) \implies 2\cdot\cos((A+B)/2)\cdot\cos((A-B)/2)$$
can be applied to the equation
$$\cos(3\cdot x) + \cos(x) = 0$$
to produce
$$2\cdot\cos(2\cdot x)\cdot\cos(x) = 0,$$
and Factorization can now be used. However, the above rule can also be applied to the equation
$$\cos(x) + 2\cdot\cos(2\cdot x) + \cos(3\cdot x) = 0,$$
to give
$$2\cdot\cos(x)\cdot\cos(2\cdot x) + 2\cdot\cos(2\cdot x) = 0$$
and Factorization can not be applied to the latter equation. (Factorization Preparation can be, of course.)

Thus, the above step is not an application of Factorization Preparation as it does not meet the requirement that Factorization must be applicable after the method has been applied. Note that step above does not satisfy the common subterms precondition of Factorization Preparation.

Another technique that can be used to make Factorization applicable involves taking logs. Taking logs base 2 transforms the equation
$$\left(2^{\cos(x)}\right)^x = 1$$
into
$$x\cdot(2\cdot\cos(x)) = 0.$$
As in the case above, the application of this technique does not always allow Factorization to be applied (if the right-hand side is a non-zero number for example) and thus isn't Factorization Preparation. Again, the application above does not satisfy the common subterms precondition.[11] We could weaken the insistence that Factorization must be applicable after Factorization Preparation, and group all three techniques into a more general method. This new method would contain rules that make Factorization *more likely* to apply, in the same way as Attraction makes Collection more likely to apply. The current version of LP does not do this. The problem with creating such a method is that the preconditions would have to be rather general, i.e. the right-hand side can be 0 or 1, the equation can be exponential or a sum etc.

2.3.1.9 Trigonometric Factorization

PRESS contains a set of specialized methods for trigonometric equations. These methods are restricted to equations consisting of linear functions of sines and cosines, equated to 0.

These methods are important for PRESS, as such questions occur often in the A level problem test set. They are also important in the context of LP, as LP was first designed for problems of this kind. For this reason, LP does not contain these methods, it attempts to learn them. This section describes

---

[11]This step is a special case of the Logarithmic method, described in section I.4.

one of the specialists, the **Arithmetic Progression Specialist**, the other specialists are described in appendix I.

The Arithmetic Progression Case

This method is restricted to equations with three trigonometric terms. The equation must be of the form
a·cos(p) + b·cos(q) + a·cos(r) = 0

or

a·sin(p) + b·sin(q) + a·sin(r) = 0.

The coefficients a, b and c are free of the unknown. The method first checks to see if the angles meet the requirement that q = (p + r)/2.[12]

Note that these conditions are very restrictive, equations that satisfy these conditions are very special cases.[13] Why implement a method to solve such equations? The answer is that A level papers usually contain a question of this form, the examiners seem to like testing whether students know the "trick" for solving this very small class of equations. If we included such special methods for every class of equation, PRESS would become a bag of such special-purpose tricks. The prevention of this is part of the motivation behind LP - let the program learn new specialist methods itself from more general methods.

Nevertheless, the arithmetic progression method has been implemented in PRESS, as such equations occur frequently in the test set.

If the requirements above are met, the first and third terms are added using the rewrite rules
sin(X) + sin(Y) ==> 2·sin((X + Y)/2)·cos((X - Y)/2)

cos(X) + cos(Y) ==> 2·cos((X + Y)/2)·cos((X - Y)/2)

As Y = (X + Z)/2, this transforms the equations to
2·a·sin(q)·cos((p - r)/2) + b·sin(q) = 0

or

2·a·cos(q)·cos((p - r)/2) + b·cos(q) = 0.

Now Factorization Preparation[14] can be applied. The equations become
sin(q)·(2·a·cos((p - r)/2) + b) = 0

cos(q)·(2·a·cos((p - r)/2) + b) = 0.

The equations can now be split into factors and solved.

---

[12] If a and b are not equal, there is no problem in deciding which angle in the problem must correspond to q, the two terms with the same coefficient have angles p and r (the order being immaterial) and the remaining term has angle q. If a and b are equal PRESS looks for the largest and smallest angles, the remaining one is q.

[13] However, more complex versions of such equations are fairly common in Physics. They are truncated fourier series.

[14] PRESS uses Collection here, Factorization Preparation is not currently a PRESS method. See section 2.3.1.8 above.

A typical equation of this kind in the test set is
cos(x) + 2·cos(2·x) + cos(3·x) = 0 (A.E.B. 1972)

Another equation in the test set,
sin(x) + sin(2·x) = sin(3·x) (London 1976)
meets many of the requirements of the arithmetic progression case. However, the sin(3·x) has a coefficient of -1 when the equation is rearranged. If the first and third terms are added, we obtain
-2·cos(2·x)·sin(x) + sin(2·x) = 0.
Now Factorization Preparation can not occur, and thus Factorization can not be applied. The equation can be solved by Homogenization (see chapter 3).

2.3.1.10 Other Methods

PRESS has three other equation solving methods. The most important of these, Homogenization, mentioned above, is discussed in chapter 3. The other two, Logarithmic Methods and Nasty Function Methods, are described in appendix I.

2.3.2 Primary and Secondary Methods

It should be observed that only two of the methods, Isolation and Polynomial Methods, actually produce solutions. These methods are called the **primary methods**. The other methods transform equations so that primary methods can be used. These methods are called **secondary methods**.

The methods can be classified along other dimensions as well. For example, Factorization and Change of Unknown increase the number of subgoals, by producing more, but hopefully simpler, equations. The other methods do not do this.

2.3.3 The Heuristic Waterfall

This section describes how PRESS uses the methods to solve equations.[15] The technique used by PRESS is similar to that used in the Boyer-Moore theorem prover. In [7], Boyer and Moore use the term **waterfall** to describe their top-level process. We will follow this terminology, and use the term **heuristic waterfall** to describe the control flow of PRESS.

The waterfall consists of a number of methods. At the top of the waterfall, PRESS checks to see if the equation is already solved. If it is, PRESS returns the answer and the equation is removed from the waterfall. Otherwise, the equation is passed over the waterfall. On the way down, the PRESS methods try to transform the equation. If a method succeeds in transforming the equation, the new equation is sent to the top of the waterfall and the process is repeated. If a method such as Change of Unknown creates more than one equation, all such equations are sent to the top. If a method fails to transform the equation, the equation falls to the next level where the next method is tried. The process terminates with success when there are no more equations to be processed. If an equation falls right through the waterfall, i.e. no method can transform the equation, PRESS backtracks. This process is described below in section 2.3.3.1. Finally, if all possibilities have been tried, and equations still remain in the waterfall, the process terminates with failure, i.e. PRESS fails to solve the equation.

PRESS tries the methods in the order Isolation, Factorization, Polynomial Methods, Change of

---

[15]The same technique is used by LP before learning has taken place, but learning modifies the behaviour, see section 4.5.

Unknown, Collection, Attraction, Trigonometric Methods, Logarithmic Methods, Homogenization and Nasty Function Methods. Generally, methods that lead directly to a solution (e.g. Isolation), or have other desirable properties (e.g. Factorization)[16], should be attempted first, and this is reflected in the ordering. The rest of the ordering was determined experimentally. Various different orderings were tried on some test problems, and the output was evaluated on several criteria. These criteria included considerations on efficiency, and whether the output resembled the output that we would produce when solving the problems ourselves.

The above description of the waterfall is slightly simplified. The process is rather more complex, due to efficiency considerations. For example, if the equation is a polynomial, and the Polynomial Methods cannot solve it, the process terminates with failure immediately. This is because none of the other methods can transform polynomials in a useful way, so the program should not bother to try. Similar remarks apply to equations that appear suitable for Factorization.

This ordering can be viewed as implementing a kind of plan. For example, the Basic Method, described in section 2.3.1.4, tries to Isolate the equation. If this is not possible, it attempts Collection. If Collection succeeds, it then attempts to apply Isolation to the new equation. If Collection fails, Attraction is attempted, followed by Collection etc. The waterfall structure and the method ordering implements this type of plan.

Note that the waterfall is a very simple control mechanism. We are able to use such a simple device because the meta-level search space is small and well-behaved. By well-behaved, we mean that if a method is applicable, it will usually be right to apply it, and if this decision is wrong, dead ends are quickly reached. The object-level search space does not have these desirable properties. Thus, the use of meta-level inference, to transform the search space from the object-level to the meta-level, transforms an ill-behaved search space to a well-behaved one.

However, the meta-level space is still too large to search it all. Some pruning is needed, the next section describes how this is done.

2.3.3.1 Pruning the Meta-Level Search Space

The waterfall selects the first applicable method, the one that comes highest in the priority ordering. Methods lower down will not usually be tried. It may be the case that some of other methods were also applicable. The role of the priority ordering is to make it likely that the method chosen is better than the other possible methods. Generally, the waterfall enforces the choice, and all the other possibilities are ignored. In these cases, the view is that the priority ordering is correct, and all the other methods are inferior. In effect, PRESS decides not to examine the part of the meta-level search-space that lies on the branches corresponding to the application of the other methods. This prunes the meta-level search-space considerably.

In other cases, the choice of method is not so certain. PRESS still tries the highest priority method first, but if this leads to an equation that cannot be solved (it falls right through the waterfall), it is worth examining other portions of the meta-level search space. To do this PRESS **backtracks**,[17] and tries an alternative method. The first choice might have eventually produced several equations in the waterfall, and the backtracking must remove all such equations.

Note that backtracking is an artifact caused by the serial nature of the machine, and the depth-first

---

[16]In the context of LP, such methods are called **key methods**, see section 4.3.2.3.

[17]The backtracking behaviour of PRESS is handled by the standard Prolog system, [23, 5].

search strategy of Prolog. On a parallel machine, the pruning of the meta-level search space may still be necessary, but where more than one branch remained to be searched, the competing branches could be examined in parallel. Similarly, a breadth-first search strategy would avoid the need for explicit backtracking, at the cost of increasing the amount of space needed by the program.

Backtracking is allowed after:

- some of the Trigonometric Factorization Methods (see section I.3),

- some of the Nasty Function Methods (section I.2),

- and two subcases of Homogenization, the mixed case (section 3.1.3.4), and the trigonometric case (section 3.1.3.5).

We could allow backtracking after **all** methods, but this produces an inefficient program which may end up searching far too much of the entire space. (It may also use up too much machine space.) We therefore try to limit the number of methods that admit choices. The decision as to which methods these should be is partly a heuristic decision, which depends on the nature of the problems that the program is expected to solve.

Why allow backtracking for the methods listed above? For the first two cases, the reason is that we have less confidence in the relative ordering of methods lower down the priority ordering. We are less certain that the first choice method is the only method that should be applied, perhaps methods lower down should also be considered.

The reasons in the case of Homogenization are slightly different. Here, various choices are made *within* Homogenization, and the program may need to examine the equations formed by different choices. (Backtracking is also needed to allow the Nasty Function Methods to be tried.[18]) If one choice in Homogenization leads to an equation that PRESS cannot solve, the program backtracks and makes another choice if possible. If all the possible choices in Homogenization lead to equations that cannot be solved, PRESS eventually tries the Nasty Function Methods. The choices within Homogenization are very heuristic, and the heuristics aren't perfect. Some examples require one choice, other examples need another, and the heuristics cannot distinguish between the two types. This is why backtracking is needed.

To recap, these decisions are a heuristic compromise. If backtracking was allowed after all the methods, PRESS would be a rather inefficient program that might sometimes need to search almost all the meta-level search space. If backtracking was not allowed at all, PRESS would be prevented from solving some equations that it can solve, which rely on the ability to undo abortive first attempts.

2.3.4 Example Equations

Figure 2-4 shows some typical equations that the current version of PRESS can solve. (PRESS can also solve all the equations shown in figure 1-1 in chapter 1). The timing information is for PRESS running on a Dec-10 KL.

---

[18] The Nasty Function Methods are the only methods below Homogenization in the priority ordering.

Equation	Time (ms)	Source
$9^{3 \cdot x^2} = 27^{15-x}$	1163	A.E.B. 1973
$(1-\tan(x)) \cdot (1+\sin(2 \cdot x)) = 1 + \tan(x)$	4089	A.E.B. 1975
$x^4 - 7 \cdot x^3 + 14 \cdot x^2 - 7 \cdot x + 1 = 0$	1229	London 1977
$\sec(x) - 1/\sec(x) = \sin(x)$	3390	Oxford 1977
$\log_x 2 \cdot \log_x 3 = 5$	1914	London 1981

Figure 2-4: More Equations That Can Be Solved By PRESS

2.4 Summary of the Methods of PRESS

Table 2-1 contains a quick summary of the PRESS methods described in this chapter. (Table I-2 in appendix I summarizes the remaining methods). In the preconditions column, "multiple occurrences" means that there must be at least two occurrences of the unknown in the equation.[19]

2.5 Discussion

This chapter has shown that the techniques used by PRESS enable it to solve equations without suffering from the combinatorial explosion. Syntactic properties of the equation are used to decide which rewrite rules should be applied next.

The rewrite rules are partitioned into sets according to the way they affect the syntactic characterizations of equation, e.g. Collection rules reduce the number of occurrences of the unknown. The fact that each set of rules should be used only in appropriate circumstances, gives rise to the concept of methods. These consist of a set of rewrite rules plus the meta-level control information indicating when the rules should be used.

Isolation, for example, consists of rules that reduce the depth of the unknown in the equation, and the control information that indicate that the Isolation rules should only be applied if the equation contains exactly one occurrence of the unknown.

The syntactic properties used by PRESS are **meta-level** characteristics. The concept of the number of occurrences of the unknown is not one of the theory of algebra, but is in the theory of the representation of algebra, i.e. it is in the **meta-theory** of algebra.

2.5.1 Discussion of Meta-level inference in PRESS

We can now continue the discussion of meta-level inference in PRESS. As we have seen, there are two distinct levels of PRESS. The first is the **object-level**. This contains the basic axioms of algebra, such as rewrite rules. The other level is the **meta-level**. This level contains the axioms of the *meta-theory* of algebra. In the meta-theory, the objects that are manipulated are algebraic, e.g.

---

[19] See section 4.3.2.1 for a discussion of preconditions.

Method	Preconditions	Description
Isolation	Single occurrence of the unknown.	Solves the equation by stripping off outer functions.
Collection	Multiple occurrences.	Reduces number of occurrences of the unknown. Applies to least-dominating subterms containing marked variables.
Attraction	Multiple occurrences.	Reduces size of least-covering tree. Applies to least-dominating subterms containing marked variables.
Factorization	Equation must be in the form $A \cdot B = 0$.	Solves each factor separately, producing new subgoals.
Polynomial	Equation must be a polynomial.	Solves the equation if possible.
Change of Unknown	Multiple occurrences, all occurrences in identical subterms.	Changes the unknown, producing two new subgoals.
Factorization Preparation (LP method)	Equation is of form $A + B + \ldots + M = 0$ where $A, B, \ldots, M$ have a common subterm containing x.	Transforms equation so that Factorization can apply.

Table 2-1: Characteristics of the Methods Described in this Chapter

variables formulae etc. As in any meta-theory, the object-level entities are considered as "marks on paper", and can be manipulated syntactically according to a set of meta-theory axioms. One such axiom, from [21], is

$$\text{singleocc}(X, L=R) \ \& \ \text{position}(X,L,P) \ \& \ \text{isolate}(P, L=R, \text{Ans}) \rightarrow$$
$$\text{solve}(L=R, X, \text{Ans}). \tag{xxi}$$

This can be considered as a procedure, to solve the goal on the right of the arrow, satisfy the subgoals on the left. However, it also has a **declarative** meaning in the meta-theory. The declarative meaning is

"If $L = R$ contains exactly one occurrence of X, and the position of this occurrence in L is P, and if the result of isolating X in $L = R$ is Ans, then Ans is a solution to the equation $L = R$, with X as the unknown."

Note that the description refers to properties such as position and the number of occurrences of X. These are meta-theory syntactic features. The description can be viewed as the statement of the correctness of the Isolation method, if Isolation can be applied, then the result produced is a solution to the original equation. The IMPRESS theorem-prover, [16, 75], was originally designed to prove theorems such as (xxi), given the definitions of isolate and position etc.

The inference of PRESS occurs at the meta-level. Some of the meta-level predicates are of the form
> New is the result of applying Rule to Old.

To satisfy such a predicate, the rule Rule is applied to the expression Old to produce New. As a result of this, an algebraic transformation has occurred at the object-level, the expression Old has been transformed into New. Thus, in PRESS, search at the object-level is replaced by search at the meta-level. This works well because the meta-level search space is much better behaved than the object-level one. In particular, the branching rate is much lower, and most wrong choices lead to dead ends rapidly. (See section 2.5.2 below for more details of the search space.) The control process of this level is trivial, and is handled by the Prolog interpreter, which provides depth first search. (See [23, 5] for details.)

If the meta-level space is complex, it is possible to axiomatize the control of *this* level, i.e. to produce a meta-meta-level. This process can continue until the control process of the current top level becomes trivial. This usually happens very early, so only two levels are needed. Thus PRESS has the object-level and the meta-level. IMPRESS also has only two levels, the meta-theory of algebra and the meta-meta-theory.

2.5.2 The Search Space

What is the search space for PRESS, and what steps does the technique of meta-level inference allow it to avoid? Consider the example discussed above, the solution of the equation
$$(2^{x^2})^{x^3} = e.$$
Figure 2-3 above shows how PRESS solves this equation, applying Attraction, Collection, and two applications of Isolation. What other steps could have been tried instead?

Starting with the original equation, many useless steps apply. Both sides of the equation could be multiplied by 1, or 0 could be added to both sides. The steps use the rules
$$X \Longrightarrow X \cdot 1$$
$$X \Longrightarrow X + 0$$
Equally, each individual term could be similarly treated, e.g. the exponent 3 could be replaced by $1 \cdot 3$ or $3 + 0$. These steps are useless for A level equation solving, so PRESS avoids these steps by not including these rules.[20]

Not all bad steps can be avoided so easily. PRESS contains many rules that would cause problems if applied inappropriately. For example, following the example in [9], we could use functional reflexive axioms, such as
$$A = B \Longrightarrow A^N = B^N.$$
So, applying this rule with $N = 1/x^3$ to the equation produces
$$2^{x^2} = e^{1/x^3},$$
after some simplification. Recalling that there is such a functional reflexive axiom for every function in PRESS, such examples suggest that the search space could be huge.

---

[20] The rules are used the other way round in the simplifier, see [76, 21] for details.

Note that the step above seems plausible. It reduces the complexity of the left-hand side, but increases the complexity of the right-hand side. The use of meta-level inference prevents PRESS from applying such rules in inappropriate situations.[21] In this case, N shouldn't contain x.

Other steps are better. Consider the use of another functional reflexive axiom:[22]
$$A = B \implies \log_2 A = \log_2 B.$$

This transforms the equation above into
$$(x^2)^{x^3} = \log_2 e.$$
This new equation can be solved by Attraction followed by Polynomial Methods.

Finding the correct rule is only part of the problem, the rule must also be applied correctly. For example, as was seen in section 2.3.1.3, PRESS contains the Attraction rules:
$$U^{(V \cdot W)} \implies (U^V)^W$$
$$(U^V)^W \implies U^{(V \cdot W)}$$
To avoid looping, Attraction applies these rules only when the terms corresponding to U and V contain the unknown and the other terms do not. Once one of the rules has been used, the other rule cannot be applied, as these conditions aren't met. Without the use of meta-level inference to apply the rules selectively, either one of the rules would have to be omitted, or some form of loop checker would have to be used.

Given that meta-level inference constrains the object-level search space, what of the meta-level search space? This is the search that PRESS as a program has to do. PRESS contains between 15 and 20 methods, counting each Nasty Function and Trigonometric specialist separately. Each method has various preconditions. Usually, most of the methods will not have all of their preconditions satisfied, so most of the methods will not be applicable. On average, probably not more than five methods have all their preconditions satisfied. Many of these methods will fail when attempted, because the rewrite rules of the method can not be applied. In our experience, the meta-level search space is usually very small. Often, wrong decisions quickly lead to a dead end. When more than one method can be applied, it is often the case that any of these methods will do, i.e. the equation can be solved using any of them, although the various solution methods may differ in efficiency.[23] This is an example of what Kowalski calls "don't-care non-determinism", [42].

2.5.2.1 The Partitioning of the Rewrite Rules

Section 2.2.2 discussed the advantages of the selective application of rewrite rules over exhaustive rewriting. A key part of selective application in PRESS is the partitioning of the rewrite rules into sets.

Note that the way the rules are partitioned is entirely dependent on the syntactic characterization used. PRESS focuses on concepts such as the occurrences of the unknown in the equation. Subsidiary concepts such as depth are also used. This allows us to partition rewrite rules into sets that reduce

---

[21]This functional reflexive rule is used in Isolation (and its derivatives, see appendix I) to remove dominating exponential functions.

[22]PRESS can perform this step using one of two methods described in appendix I, Function Stripping or the Logarithmic method.

[23]This is not always true. See equation (iii) in section 3.1.6 for an example of an exception.

the number of occurrences, decrease the depth, etc.

A different classification might give different results. An obvious example is to divide the set into rules that deal with trigonometric equations, logarithmic equations etc. In fact, usually pattern-matching will provide this effect anyway. For example, the rule
$$\log_X Y + \log_X Z \implies \log_X Y \cdot Z$$
can only be applied to logarithmic equations.

A problem with this type of classification is that there are many rules that are useful for many types of equations, for example, the Collection rule
$$M \cdot X + N \cdot X \implies (M + N) \cdot X.$$
Several versions of this would be needed, one for the cases in which X is a trigonometric term, one when X it is a logarithmic term etc, or worse, a version of the rule one may be needed for each individual subtype, e.g. sin, cos etc. This creates problems, as the "classification" does not divide into disjoint classes, rules may combine terms from different types. There would have to be a copy of each rule for each type.

This particular classification does not appear to have advantages over the existing one.[24] However, this does not mean that no better classification exists, we just do not know of one!

2.5.2.2 The Control Information

The control information also uses concepts from the syntactic characterization of the equation, and are similarly dependent on the choice of characterization used.

In general, the control information includes a recognizer, which indicates whether an equation is in the domain of applicability of the method, and a place in the **priority ordering**. The recognizers usually check necessary conditions rather than sufficient ones. For this reason, different methods may have the same recognizer, for example Collection and Attraction both check that the equation contains at least two occurrences of the unknown.

The priority ordering decides the order in which methods should be tried. If methods have unique recognizers this is not important. In theory, if methods share recognizers, as is the case for Collection and Attraction, the order still does not matter, as bad decisions can be undone by backtracking. However, this ignores efficiency considerations, and the ordering can be crucial for efficiency, computationally cheap methods should be tried before expensive ones. If two methods have the same computational cost, the more powerful method should be tried first. For example, Collection and Attraction have roughly the same cost, but Collection is the more powerful method. PRESS tries to use Collection before Attraction.

2.5.3 Adding New Methods

Chapter 3 and appendix I describe methods that were added to PRESS during the course of its development. These methods must integrate with existing methods. To do this each new method must contain a recognizer, rewrite rules, and, perhaps, a position in the priority ordering. LP, described in chapter 4, creates new methods which must be added to its existing ones. The problem is discussed in section 4.4.2.1.

---

[24] In fact, the two are complementary to some extent, e.g. the Collection rules could be subdivided into logarithmic Collection rules, trigonometric Collection rules, general Collection rules etc.

2.5.3.1 The Performance of PRESS

PRESS performs well as an equation solving program. It currently solves 86% of the A level test set, a figure comparable to very good A level students. Why not 100%? To answer this question, we need to consider the contribution of the new specialized methods, described in chapter 3 and appendix I.

Much of the figure of 86% is due to the new powerful methods, Homogenization and the specialists described in appendix I. Without these methods PRESS would be able to solve only about 35% of the problems. Against this, the new methods increase the number of PRESS methods by 25-50%.[25] The introduction of these methods was largely problem driven. At any stage, there are problems in the test set that PRESS cannot solve. It is easy to write code that will enable PRESS to solve any one particular problem, but such an approach is uninteresting, and possibly harmful, as code for one equation may interfere with code for another.

To justify inclusion as a PRESS method, an equation solving technique should be general enough to solve a range of problems, and such problems must occur reasonably often in the test set. The terms "range" and "reasonably often" are not precisely defined, (although they could be arbitrarily defined if this was desirable), the decision is one for the programmers.

Many of the unsolved questions require specialized techniques. More new specialized methods would be required to get PRESS to solve these remaining questions. These new methods would be motivated by relatively few problems, and the situation may be reached where a new method is required for just one problem. Such a method obviously does not meet the requirements above.

An analysis of the remaining unsolved problems showed that about six new methods might be sufficient to allow PRESS to solve the entire problem set. This would represent a 50% increase in the number of PRESS methods, and the problems on the next new exam paper might require still more new methods.

The modular waterfall design of PRESS (see section 2.3.3) minimizes the effort involved in implementing a new method. The major problems involve undesirable interactions with other methods[26]. Nevertheless, a fair amount of time is required to write, debug, and integrate a new method.

While it is certainly possible to add six new methods to PRESS, an alternative approach is to build a learning system that would add new methods itself, integrating them with its existing methods. This is one of the motivations of LP, described in chapter 4.

2.6 MACSYMA

How does PRESS compare with other similar programs? Perhaps the most famous algebraic manipulation systems is **MACSYMA**, [49]. This section compares PRESS with MACSYMA, many of the comments here apply to other similar systems such as **REDUCE**, [39].

If only equation solving is considered, PRESS easily outperforms MACSYMA. For example, MACSYMA can only solve equations using versions of Isolation, Factorization, Change of Unknown

---

[25] The exact figure depends on whether a set of methods, such as the Nasty Function Methods is counted as one method or several.

[26] Of course, the ability of PRESS to solve many equations owes a lot to desirable versions of these interactions!

and Polynomial methods. MACSYMA is equally restricted on simultaneous equations. PRESS can solve sets of non-linear equations as well as linear ones (see section 3.2), whereas MACSYMA can only solve the linear case. However, MACSYMA was not intended to be an equation solving program only. It is a general algebraic manipulation program.

MACSYMA is produced and maintained by the Mathlab Group of the MIT Laboratory for Computer Science. It is written in LISP. Apart from equation solving, it can differentiate, integrate, take limits, factor polynomials, expand functions in Laurent or Taylor series, solve differential equations, compute Poisson series, plot curves, and manipulate matrices and tensors.

Of these other skills, PRESS can only differentiate and integrate (see [71]).

Search control does not appear to be a large problem for MACSYMA. This is because MACSYMA is a very **interactive** system. In general the user has to tell it what is should do next. When solving polynomials or using Isolation etc, MACSYMA uses algorithms. However, it is interesting to note that MACSYMA performs very little checking. For example, if it is given a set of non-linear equations, it will assume that they are linear and get the answer wrong.

In summary, PRESS and MACSYMA differ in many respects. MACSYMA can do many more kinds of algebraic manipulation than PRESS can. However, where their abilities overlap, equation solving, PRESS outperforms MACSYMA.

2.7 The Psychological Validity of PRESS

As has been previously stated, we do not wish to claim that PRESS has great psychological validity, i.e. we do not say that PRESS is an accurate model of human equation solving.

However, some parts of PRESS may have some psychological plausibility, whereas other parts certainly do not. All the evidence is very subjective. One crude way of evaluating the psychological validity of PRESS is to introspect about our own equation solving behaviour and compare it with the behaviour of PRESS. However, this is unsatisfactory in two ways. Firstly, some of the key ideas for PRESS came in the first place from such introspection. Secondly, those of us who have worked on PRESS for a long time have often modified our own equation solving behaviour, e.g. the author now solves equations using PRESS methods!

We must therefore rely on the judgements of other people. Several skilled equation solvers have examined the output trace produced by PRESS, and have often agreed that they would produce a similar trace on the problems. Generally, people feel that the meta-level behaviour, i.e. the methods and the waterfall, has the greatest psychological validity.

Different people may have different sets of methods, but most skilled solvers have the more general methods, such as Collection.

As mentioned above, in section 2.5.2.1, a different partitioning of the rules would lead to different methods. Thus human solvers may think more in terms of methods to solve trigonometric equations, methods to solve logarithmic equations etc. However, given a trigonometric equation, the solver would consider applying those rules which reduce the number of occurrences of the unknown, i.e. the trigonometric Collection rules.

As far as the waterfall is concerned, the priority ordering differs from person to person. Even so, it would indeed be surprising if a person did not try a version of Factorization or Isolation until after some of the Nasty Function Methods say. Many people, on being shown the "basic method" of applying Attraction, Collection and Isolation, feel that this closely mirrors their own "basic plan".

Parts that lack psychological plausibility include some of the more "black-box" sections, such as the polynomial package and the tidying routines. The former uses a polynomial normal form, that is suitable for a computer but not really useful for a human solver. Both these systems perform extensive numerical calculations to produce a result, whereas a human would split the problem into many more steps.

Similar comments apply to some of the Homogenization routines, described in chapter 3 and appendix II.

Some psychologists have expressed interest in using PRESS as a model of the human equation solving process, [25, 22]. See [22] for more details of the most extensive study.

Separate but related to the question of psychological validity is the issue of the educational value of PRESS. PRESS could prove to be of pedagogical value in at least two ways. Firstly, a version of PRESS could be used as the basis of a computerized teaching system. At present, no work is being done in this direction. Secondly, students could be taught to solve equations using the concepts and methods of PRESS, (psychological validity in reverse!) Ron Wenger of the University of Delaware is considering this possibility.[27]

2.8 Conclusions

This chapter has described the PRESS program and the technique of meta-level inference. Meta-level inference involves two levels: the object-level that encodes factual knowledge about the domain, and the meta-level that encodes the control are strategic knowledge.

The control knowledge in the meta-level is expressed as axioms in the meta-theory of the domain. Inference at the meta-level causes inference at the object-level.

The advantages of this technique are:

- The separation of the factual and control information enhances the clarity of the program and makes it more modular.

- The search process can be controlled in a flexible way. In particular, the search process is data-sensitive.

- Generally, the meta-level search space is smaller than that of the object level, and this reduces the problems of combinatorial explosions.

2.8.1 PRESS and LP

This chapter has provided a general description of PRESS. The remaining methods of PRESS are described in chapter 3 and appendix I. However, none of this additional material is needed to follow the description of LP in chapter 4, and this chapter can be read next if desired.

---

[27]Personal communication.

CHAPTER 3

HOMOGENIZATION

This chapter describes in detail a powerful, general purpose method, **Homogenization**. Homogenization is a more complex method than those described in chapter 2. The search space is huge, but Homogenization works in a very directed way, very little search is needed.

The first section describes Standard Homogenization, Homogenization applied to equations in one unknown. The next section describes Extended Homogenization, Homogenization applied to solve simultaneous equations.

Section 3.3 discusses how Homogenization can be useful in domains other than equation solving.

Origins

The idea of Homogenization was first suggested by Alan Bundy in [10]. However, the description in the paper was an outline, omitting many important details. For example, the paper gives only vague suggestions as to how the reduced term should be selected. The actual implementation of Homogenization differs in many ways from the description in [10].

3.1 Standard Homogenization

In chapter 2 we saw that the Change of Unknown method is a potentially useful equation solving method. However, it relies on the unknown occurring within identical subterms in the original equation, and usually equations do not have this property.

Consider the example of chapter 2. In the equation

$$4 \cdot \log_x 2 + \log_2 x = 5 \qquad \text{(London 1978)} \tag{i}$$

the occurrences of the unknown, x, appear within dissimilar subterms, namely $\log_x 2$ and $\log_2 x$. It therefore requires some preparation before Change of Unknown can be applied. In particular, one of the subterms $\log_x 2$ and $\log_2 x$ must first be converted to the other, with the aid of the rewrite rule

$$\log_U V \implies 1/\log_V U.$$

This preparation step is called **Homogenization**, because it makes the occurrences of x appear in identical subterms, and hence makes the equation more homogeneous in x.

3.1.1 Some Terminology

Before the Homogenization Method can be described, some terminology must be introduced.

- The original equation, prior to Homogenization, will be called the **input equation**, and the resulting equation will be called the **output equation**.

- The output equation belongs to a class of equation which will be called the **base class**. The first limitation is to restrict the base class to **algebraic** equations.

  Algebraic equations are those involving only the functions +, -, ·, / and exponentiation to a rational number. Thus
  $$((x^{1/2} + 1)/(2 \cdot x - 1))^{1/3} = 3$$
  is algebraic in x, whereas $\cos(x) = 1/2$ is not, as it involves the transcendental function cos.

  The input equation can always be regarded as an algebraic equation *in some set of non-algebraic subterms in x*, e.g. equation (i) above can be regarded as an algebraic equation in the set $\{\log_x 2, \log_2 x\}$. These subterms are called the **offending terms** and the set of them is called the **offenders set**. The idea is that the offenders set is a set of subterms that prevent the equation being algebraic: a type of equation which PRESS knows a lot about.

  The output equation is algebraic in a single non-algebraic subterm of x, f(x). Then substituting y for f(x) produces an equation that is algebraic in y.[1]

- The essence of the Homogenization method is to replace each of the offending terms by some algebraic function of a single term, called the **reduced term**. In the example above the reduced term is $\log_2 x$.

Note that the output equation has two properties. Firstly, all the occurrences of the unknown are in identical subterms, so that the Change of Unknown method can be applied. The second property is that the output equation is algebraic in *one* non-algebraic subterm, so that once Change of Unknown has been applied, the changed variable equation is algebraic in the new variable. This strategy is adopted because PRESS is good at solving algebraic equations, so it is quite likely that the changed variable equation can be solved.

Both of these conditions are meta-level properties, i.e. they are syntactic properties of the output equation.

3.1.2 The Homogenization Method

Viewed declaratively, Homogenization establishes a relation:
homogenized(X,Input,Output,Reduced)
between the input equation Input, the output equation Output, the reduced term Reduced and the unknown X, such that:

(i) Input and Output are equivalent, i.e. either can be obtained from the other by legal algebraic moves.

---

[1] All the examples that we have found in text books and exam papers have output equations of this form. In fact, most output equations are (sometimes disguised) quadratics in f(x). Any class of equations could be used as the base class, but the idea is to pick a class whose members are relatively easy to solve.

(ii) Output is algebraic in a single non-algebraic subterm, Reduced, containing X.

In more detail, the Homogenization method is axiomatized as

offenders_set(X,Input,Set,Type) &
reduced_term(X,Input,Set,Type,Reduced) &
rewritten(Input,Type,Set,Reduced,Output) →

$\quad$ homogenized(X,Input,Output,Reduced) $\hfill$ (ii)

where Set is the offenders set, of type Type (see below).

As described in section 2.5.1 above, PRESS performs inference at the meta-level. When it attempts to satisfy axiom (ii), some moves are induced at the object-level, e.g. eventually the equation Output is constructed. The usual advantages of meta-level inference apply to the method of Homogenization. In particular, there is no need for the programmer to consider every possibility, the fairly simple meta-level axiom (ii) can induce arbitrarily complex object-level transformations if necessary.

Homogenization can also be viewed procedurally. Algorithmically, the Homogenization method is as follows:

(a) The offenders set is found by trying to parse the input equation as an algebraic equation. The parse is blocked when the current subterm is x or a non-algebraic subterm containing x. This term is then added to the offenders set[2] and the parse is forced to continue.

(b) If the offenders set is a singleton, three cases must be distinguished.

$\quad$ (i) The singleton is the unknown, x. In this case, the equation is already algebraic, so Homogenization should not be tried. Exit with failure.

$\quad$ (ii) The singleton is a function of x, (say f(x)) which occurs at least twice in the input equation. Exit with success, as now the Change of Unknown method can be applied to the output equation, substituting y for f(x).

$\quad$ (iii) The singleton is a function of x as above, but occurs only once in the equation. In this case the Change of Unknown method cannot be used so exit with failure. This case includes examples like
$$x^{\cos(x)} = 2,$$
where the offenders set is the singleton
$$\{x^{\cos(x)}\}.$$

(c) Classify the offenders set according to the type of its members. For example, if all the offenders are trigonometric terms, then the set is of type trigonometric. The classification scheme is explained in section 3.1.3.2 below.

(d) A reduced term is selected. The type of the offenders set determines how this term is chosen, different procedures are needed for trigonometric terms than for exponential terms, for example. In exceptional circumstances, where the special methods fail, the reduced term is chosen from the offenders set on the basis of a simplicity measure.

---

[2] As the offenders set is a *set*, each element appears only once, so addition here means set union.

(e) Now an attempt is made to rewrite each term in the offenders set as an algebraic function of the reduced term. In some cases the transformation needs only one rewrite rule, but other terms will require a series of rewrites.[3] If the method succeeds a rewrite is found for every term in the offenders set, i.e. each offending term o_i is rewritten as an algebraic function f_i of the reduced term t. If no rewrite rule is applicable, then backtrack to choose a new reduced term if this is possible, otherwise fail.

(f) Substitute the rewrites for the offending terms in the input equation to give the output equation. This equation is now an algebraic equation of the reduced term, i.e. it is homogenized, so exit with success. Change of Unknown can now be attempted, substituting y for the reduced term in the output equation. (Note that Change of Unknown can fail if PRESS cannot solve changed variable equation or the substitution equation.)

3.1.3 Selecting the Reduced Term

The selection of the reduced term is the key step in Standard Homogenization. A wrong choice may cause Homogenization to fail. This will happen if the terms in the offenders set cannot all be rewritten as algebraic functions of the reduced term. Less seriously, a bad choice may cause the output equation to be unnecessarily complex.

The selection of the reduced term is dependent on the rewrite rules available. For example, if the offenders set contained sin(3·x) and sin(x), the selection of sin(3·x) as the reduced term would be wrong if there were no rule for rewriting sin(m·x) in terms of sin(n·x) for m < n.

Given a set of rewrite rules, the task is to ensure that a reduced term is chosen that is most likely to allow successful rewriting of the offenders set. As pointed out above, different methods are applicable to the various types of offenders sets that arise.

It had been hoped that it would be possible to choose the reduced term in a uniform way for all the problem subtypes. In order to obtain reasonable results, however, special purpose methods had to be implemented for each subtype.[4]

While these methods are highly specialized, a learning system should be able to acquire them by refining weaker, more general purpose techniques.

3.1.3.1 The Degree of the Output Equation

Before discussing the individual types, we should note another consideration that affects the choice of the reduced term. In general, it is desirable that the output equation is as simple as possible. Consider the following equation:

$e^{(4 \cdot x)} + 2 \cdot e^{(2 \cdot x)} + 1 = 0.$

---

[3] For example, one rewrite rule is of the form
(tan(A) => F(T)/G(T)) ←
    (sin(A) => F(T) & cos(A) => G(T))
where F, G are algebraic functions, T is the reduced term. This rule says that if an offending term is of the form tan(A), rewrite sin(A) and cos(A) as algebraic functions of T, calling other rules recursively if necessary. The required rewrite of tan(A) is the quotient of the 2 terms obtained.

[4] This is another example of the fact, mentioned in chapter 1, that general-purpose methods tend to be weak.

If e^x is chosen as the reduced term, Homogenization produces the equation

$$(e^x)^4 + 2 \cdot (e^x)^2 + 1 = 0.$$

Substituting y for e^x produces the quartic equation

$$y^4 + 2 \cdot y^2 + 1 = 0.$$

This is a disguised quadratic in y^2. PRESS will discover this,[5] but it will waste time trying other tests first. If instead $e^{(2 \cdot x)}$ is chosen as the reduced term, Homogenization produces the output equation

$$(e^{(2 \cdot x)})^2 + 2 \cdot e^{(2 \cdot x)} + 1 = 0,$$

which, on substituting $z = e^{(2 \cdot x)}$, becomes

$$z^2 + 2 \cdot z + 1 = 0.$$

This is now just a normal quadratic, and PRESS can solve this very easily. Note that the exponents involved in the equation are $4 \cdot x$ and $2 \cdot x$. The greatest common divisor of the coefficient of x in the exponents is 2. By using the greatest common divisor as the coefficient of x in the exponent of the reduced term, the exponents in the output equation can be made as low as possible.

3.1.3.2 The Types of the Offenders Set

The method used to pick the reduced term varies according to the type of the offenders set. Currently the following classification of sets is used:

- Exponential: All terms are of the form u^v, where v contains the unknown and u does not.

- Logarithmic: All terms are logs, with the unknown contained either in the base or the argument.

- Trigonometric: All terms in the set are trigonometric functions. (Inverse-trigonometric functions are not included.)

- Hyperbolic: All terms are hyperbolic functions.

- Mixed hyperbolics and exponential: Every term is either an exponential of the form $e^{f(x)}$, or a hyperbolic, and both types of term occur.

- Mixed: Any set that does not fall into one of the above types.

The classification depends on purely syntactic characterizations of the offenders set, and so it is a meta-level property.

The following sections describe the general-purpose method for selecting the reduced term and the method used in the case where the offenders set is trigonometric. The remainder of the special-purpose methods are described in Appendix II. The concept of the greatest common divisor occurs in most of these descriptions. The rationale in each case is the same as in the example above. Using the greatest common divisor will generally produce the lowest degree output equation.

---

[5]This equation could be turned into a quadratic by applying Homogenization again, but PRESS does not do this. Instead, the Polynomial Methods check for this disguised quadratics and other similar cases. See section 2.3.1.6 above.

3.1.3.3 The Simplicity Method

This section describes the default method for choosing the reduced term, the simplicity method. This method is used for the mixed offenders set, and is used for the other kind of sets if the special purpose methods fail.

A **simplicity metric** is applied to each term in the offenders set and the least complex term is chosen. The simplicity metric is the number of arcs in the formula tree of the term. Thus $\sin(x)$ is less complex than $\sin(x^2)$, and both are less complex than $\tan(\sin(x^2))$.

The simplicity method has the limitation that the reduced term must be in the offenders set. The other methods are not constrained in this way. For this reason the simplicity method is only used as a last resort. This means that by the time the simplicity method is used, except in the mixed case, all the special purpose techniques have failed, and in practice this usually means that the equation cannot be homogenized. However, if the simplicity method is used first, so that it processes the full range of equations, it is surprisingly successful. This was the case in earlier versions of PRESS, before the special purpose techniques were implemented. Section 3.1.5.2 discusses the performance of the simplicity method in more detail.

3.1.3.4 The Mixed Offenders Set

The process used to select the reduced term is straightforward in the case of the mixed offenders set. The simplicity method described above is used.

If a rewrite cannot be found for a term in the offenders set, then, in the case of mixed offenders sets, backtracking is allowed. If two or more terms are tied as the least complex term then they may be chosen in turn as the reduced term.

If the offenders set is mixed, Homogenization will usually fail. In earlier versions of Homogenization, the mixed set was more important, as not all the special-purpose selection procedures had been written.

3.1.3.5 The Trigonometric Offenders Set

This case of the trigonometric offenders set is the most complex. The terms in the offenders set are all trigonometric, and the first step in the method is to find the greatest common divisor (g.c.d.) of the arguments (angles) of these terms.

The method assumes that all the angles involved are linear polynomials of the form $a_i \cdot x + b_i$. When finding the g.c.d., additive constants are ignored. For example, the g.c.d. of $\{2 \cdot x + 30, 4 \cdot x + 45\}$ is $2 \cdot x$. If for some reason the g.c.d. cannot be found the simplicity method is used. See section 3.1.7 for an example of this case.

The offenders set is now further classified. It is examined to see if it is one of three following special types. These types are:

- sin-cosine pair: The offenders set contains only two terms, a sin and a cosine, both having the same argument.

- secant-tangent pair: As above but with secant and tangent in place of cosine and sin.

- cosec-cotangent pair: Again, as above but with cosecant and cotangent.

If the set does fall into one of these types, the input equation is examined to see if either of these terms occurs raised only to an even power. If this is the case, the *other* member of the pair is the reduced term. The reason for this is that each member of a pair can be expressed in terms of the other one, but the expression involves a square root, e.g.

$\sin(x) = (1 - \cos(x)^2)^{1/2}$.

However, if the term is raised to an even power no radicals occur, and usually, after Change of Unknown, a polynomial results.

As an example of the above case, consider

$7 \cdot \sin^2(x) - 5 \cdot \sin(x) + \cos^2(x) = 0$ (London 1977)

Here the offenders set is $\{\sin(x), \cos(x)\}$ so it is a sin-cos pair case. $\sin(x)$ occurs in the input equation raised to the first and second powers. $\cos(x)$ occurs only to the second power. Hence $\sin(x)$ is chosen as the reduced term.

The output equation is then

$7 \cdot \sin^2(x) - 5 \cdot \sin(x) + (1 - \sin^2(x)) = 0$

which reduces to a quadratic after substituting y for $\sin(x)$.

If, as is usually the case, the offenders set is not one of the pair forms, or both members of the pair occur at least once to an odd power, a different technique must be used. Let g be the g.c.d. of the angles in the offenders set, which was calculated as the first step.

The next step is to see if a term of the form f(g), where f is a trigonometric functor, occurs more than once in the input equation. If such a term does exist the procedure is as follows.

If f is sin or cos or tan then the reduced term is f(g). If f is cosec, sec, or cot the reduced term is 1/f(g), (i.e. sin(g) or cos(g) or tan(g)).

By choosing the most commonly occurring term as the reduced term the number of rewrites needed is reduced. This is the case for sin, cos and tan. With the reciprocal functions, the inconvenience of having to rewrite them in terms of their reciprocals is traded off against the ability to keep the number of rewrite rules small.

For example, PRESS has a rule which says to rewrite sec(y) in terms of some reduced term t, rewrite cos(y) in terms of t and invert the result. Thus this one rewrite rule replaces several explicit ones, telling how to rewrite sec(y) in terms of sin(x), cosec(x), and the other trigonometric functions.

If no term occurs more than once in the equation then what PRESS does is again governed by the structure of the rewrite rules. The trigonometric functions are given a 'niceness' order. Sin and cos are nicer than their reciprocals, which are nicer than tan and cot.

The reduced term is sin(g) if sin(g) is in the offenders set, and similarly for cos(g). If neither term occurs, sin(g) is the reduced term if cosec(g) is in the set or cos(g) if sec(g) occurs.

If the offenders set contains none of cos(g), sin(g), sec(g) or cosec(g) then sin(g) is chosen if *any* sin term occurs in the offenders set. Similarly, cos(g) is chosen as the reduced term is the offenders set does not contain a sin term, but does contain a cos term. If no sin or cos term occurs in the offenders set then sin(g) is chosen if a cosec term occurs, cos(g) if a sec term.

If a reduced term has still not been chosen, the offenders set must contain only tan or cot terms (or both), and the reduced term is tan(g).

The ordering above ensures that there is no need to express sin, cos and the other trigonometric functions in terms of tan.[6] This is desirable as the rewrites concerned are messy.

Note that Homogenization always succeeds on trigonometric offenders sets,[7] as the method of selection of the reduced term ensures that every term can be rewritten. This is not the case with other types.

However, the output equation is sometimes too difficult for the other parts of PRESS to solve. Often there are many radicals in the output equation. In the case of trigonometric offenders sets if PRESS fails to solve the output equation, Homogenization is allowed to backtrack. In this case it considers using the tan half-angle method.

This method makes use of the following rewrite rules:
$\sin(X) \Longrightarrow 2 \cdot \tan(X/2)/(1 + \tan^2(X/2))$
$\cos(X) \Longrightarrow (1 - \tan^2(X/2))/(1 + \tan^2(X/2))$

(Homogenization derives the rules for tan, cot etc, from these rules and the other trig rules.)

The method is only applicable if the angles occurring in the offenders set are either g or $2 \cdot g$, where g is the g.c.d. as above. If this is the case, and backtracking does occur, then this method is tried. It is used only as a last resort because it can introduce rational functions into an equation that was previously free of them.

3.1.4 Producing the Output Equation

Once the reduced term has been selected, Homogenization attempts to rewrite all the terms in the offenders set. Choosing which rewrite rule to use requires very little search. Most[8] Homogenization rules are of the form
C → (o ⟹ f(r)),
where C is a set of conditions, o is an offending term, r is the reduced term, and f is an algebraic function, such that o = f(r). In our implementation, each rule is indexed off both o and r, so that given an o and an r, Homogenization can quickly find the rules relating them, and in general there will be only one. Occasionally, Homogenization will have to search for the correct rewrite rule. This happens because some of the rules are recursive in nature, and several rewritings will be performed before Homogenization discovers that a wrong choice has been made. However, even these events do not cause large searches, as the number of possible rules is still very small.

Sometimes, Homogenization will be unable to find a rewrite for an offending term. If the equation is trigonometric or mixed, it may be possible to backtrack, otherwise Homogenization fails.

Suppose that a rewrite for every term has been found, i.e. for every offending term o_i there is a

---

[6] Except in the tan half-angle case described below.

[7] Provided that the g.c.d. of the angles can be calculated. If this is not possible then the simplicity method is used. This case is discussed in section 3.1.7.

[8] Some rules call other rules, producing a chain of rewrites.

rewrite $f_i(r)$, where f and r are as above. The output equation is obtained by substituting $f_i(r)$ for every occurrence of o_i in the input equation.

Now Change of Unknown can be applied, substituting a new variable for r.

3.1.5 The Search Involved in Homogenization

Despite the size of the search space, Homogenization performs very little search. Much of the method is algorithmic. The two points where search is required are the finding of the reduced term, and the rewriting of the members of the offenders set.

Once the type of the offenders set has been found, Homogenization uses one of the techniques described above and in appendix II to pick the reduced term. In some of the more complex cases, such as the trigonometric one, some search may be required to decide which subcase applies. However, all the false starts quickly come to a dead end as the program has to consider only a small number of angles and functions. In general, the search involved is trivial.

The rewriting step was discussed in section 3.1.4 above.

Experience shows that Homogenization transforms equations much more efficiently than weaker methods such as Collection (section 2.3.1.2). This is because Homogenization is very directed, the offending terms have to be rewritten as functions of the reduced term. In contrast, Collection has to try to apply a set of rewrite rules to least-dominating subterms of the equation, and it has minimal guidance to help in this task.

3.1.5.1 Results and example equations

During a survey of about 100 questions on our exam papers, about 50 questions were discovered on which the Homogenization method could be used, although some examples should be solved by better methods. All of these questions were successfully processed by the implementation of Homogenization, that is the correct output equation was found. The rest of PRESS was then able to solve 45 of these output equations. Homogenization seems to be the best method for 35 of these 45.[9]

Figure 3-1 shows some examples of equations which have been solved using Homogenization. The timing information is for PRESS running on a Dec-10 with a KL processor.

3.1.5.2 Performance of the Simplicity Method

In section 3.1.3.3 above, we remarked that the simplicity method was surprisingly effective. To quantify this, the simplicity method could be used to solve about 27 of the 35 questions best solved by Homogenization.

The simplicity method failed in 8 cases as the required reduced term was not a member of the offenders set. One such example is

$4^x - 2^{x+1} - 3 = 0$.

Here the reduced term should be 2^x, which is not in the offenders set.

Of course, as remarked above, a term is not a good reduced term per se, it depends on the rules in the rewrite rule set. The simplicity method ignores such considerations, and this can lead to a bad

---

[9]The remaining 10 problems are best solved using the trigonometric and logarithmic specialist methods, see appendix I.

Equation	Time (ms)	Source
$\cos(2 \cdot x) + 3 \cdot \sin(x) + 1 = 0$	1130	A.E.B. 1973
$3 \cdot \cos^2(x) + 5 \cdot \sin(x) - 1 = 0$	1232	A.E.B. 1975
$\log_x 8 + \log_8 x = 13/6$	1123	A.E.B. 1975
$4^x - 2^{x+1} - 3 = 0$	1096	A.E.B. 1976
$9 \cdot \cosh(x) - 6 \cdot \sinh(x) = 7$	1672	London 1976
$\sin(2 \cdot x) = \sin(x)$	776	London 1979
$\log_x 2 \cdot \log_x 3 = 5$	1979	London 1981

Figure 3-1: Some Equations that are solved by Homogenization

choice of reduced term.

In some cases, the term chosen by the simplicity method would have been a reasonable choice with a different rewrite rule set. For example, full Homogenization picks e^x as the reduced term[10] for

$2 \cdot \cosh(2 \cdot x) - 2 \cdot \sinh(x) = 3$,

and there are rules for expressing $\cosh(2 \cdot x)$ and $\sinh(x)$ in terms of e^x. However, the simplicity method selects $\sinh(x)$ as the reduced term. This is a reasonable choice[11] but the rewrite rule set does not contain a rule for expressing $\cosh(2 \cdot x)$ in terms of $\sinh(x)$.

Using the more powerful selection techniques allows the number of rewrite rules to be kept smaller, as we can ensure that we will never have to perform certain rewrites, e.g. $\cosh(2 \cdot x)$ in terms of $\sinh(x)$. In contrast, the simplicity method may require arbitrary rewrites. Even if the right reduced term is picked in one situation, a slight perturbation of the problem can make another term the most simple. The simplicity method is not very robust! What is surprising is that it works so well on the problem set.

The simplicity method chooses the same reduced term as the full implementation of Homogenization in 18 cases. In the 9 cases where a different reduced term was selected, the decision of the simplicity method was markedly worse in 8. As an example, as described in section 3.1.3.5 full Homogenization picks $\sin(x)$ as the reduced term for

$3 \cdot \cos^2(x) + 5 \cdot \sin(x) - 1 = 0$.

PRESS solves the equation in a total of 1357 milliseconds, using full Homogenization.

The offenders set is $\{\cos(x), \sin(x)\}$, and the simplicity metric chooses one of these equally simple terms arbitrarily. It so happens that the current implementation chooses $\cos(x)$. If Homogenization

---

[10] See section II.3.

[11] If the terms were trigonometric rather than hyperbolic. i.e. the offenders set was $\{\cos(2 \cdot x), \sin(x)\}$, full Homogenization chooses $\sin(x)$ as the reduced term, corresponding to the choice of $\sinh(x)$ here.

proved impossible with this reduced term, the process would back up and select sin(x), but cos(x) *can* be used. It just produces a more complex equation, and PRESS takes 4550 milliseconds to solve the equation with cos(x) as the reduced term. This 3:1 ratio seems typical of the eight cases.

It should be noted that when the simplicity method picks the same reduced term as full Homogenization, the former process is somewhat quicker, as less processing is done. This is particularly true for trigonometric equations, where the simplicity method is about 10% faster.

In summary, the simplicity method actually performs quite well on the test set. The specialist techniques are needed to solve some of the equations, and to get more efficient solutions in many cases. The specialists are also much more robust than the simplicity method, although the current problem set does not demonstrate this particularly.

3.1.6 Cases when Homogenization should not be used

As has been shown, a large variety of equations can be solved by Homogenization. However, Homogenization is unsuitable for some types of equations. There are four types of equation for which this is the case:

1. The equation does not satisfy the preconditions of Homogenization. The major precondition is that the offenders set should contain more than one element. If there is only one element, there are the three possibilities mentioned in section 3.1.2: the equation may already be algebraic, or Change of Unknown may be directly applicable, or it may be an equation that Homogenization cannot attempt, such as

    $$x^{\cos(x)} = 2.$$

    Another precondition is that the equation contains more than one occurrence of the unknown. If this is not the case, Isolation should be applied.

2. Homogenization fails on the equation, although the preconditions are satisfied. In this case, the program is unable to rewrite all the members of the offenders set as algebraic functions of the reduced term.

3. Homogenization succeeds on the equation, but the resulting equation after Change of Unknown cannot be solved.

4. Homogenization succeeds, and the resulting equation is solved, producing the correct answer. However, there is another method for solving the equation that is considered better.

In general, case 1 should not cause problems, as PRESS is arranged so that Isolation, Polynomial Methods, and Change of Unknown are tried before Homogenization.

In cases 2 and 3, the Homogenization approach is eventually unsuccessful. PRESS may still be able to solve the equation by other methods. If no other method is suitable, nothing can be done, PRESS is unable to solve the equation.

If another method is applicable, PRESS will eventually find it after Homogenization has failed. There is some inefficiency here, it would be preferable if the successful method had been tried before Homogenization. The situation is similar to that in case 4.

In case 4, as Homogenization does not fail, the other methods must be tried before Homogenization.

Examples of these cases will now be given.

A simple example of case 1 is
$\sin(x) = 1$.
This equation should obviously be solved by Isolation.

An example of case 2 is the equation
$\sin(x) + e^x = 2$.
Homogenization fails on this equation, and no other method is applicable. Therefore this equation cannot be solved by PRESS. An example of an equation where another method does exist is
$10^{x-3} = 2^{10+x}$. (London 1976)
Although Homogenization fails on this equation, the equation can be solved by the **Logarithmic Method**, described in the appendix I.

The equations, both from a 1975 A.E.B. A level paper,
$\sin(x) - \sin(4 \cdot x) + \sin(7 \cdot x) = 0$ (iii)

$\sin(3 \cdot x) = 2 \cdot \cos(2 \cdot x)$ (iv)

are examples of case 3. In both cases, Homogenization can be successfully applied. However, PRESS inefficiently transforms the output equation from (iii) to a high degree polynomial, which it fails to solve. PRESS can solve this equation using the Arithmetic Progression specialist described in section 2.3.1.9.

After Homogenization and Change of Unknown, equation (iv) gives rise to the cubic
$4 \cdot y^3 - 4 \cdot y^2 - 3 \cdot y + 2 = 0$.
This polynomial cannot be solved by PRESS, as it lacks an integer root.[12] Whereas equation (iii) can be solved with a PRESS method other than Homogenization, no PRESS method is successful equation (iv).

Finally, an example of case 4 is the equation
$\cos(x) + \cos(3 \cdot x) + \cos(5 \cdot x) = 0$ (A.E.B. 1975)
This equation can be solved using Homogenization. Homogenization and Change of Unknown transform the equation into the quintic polynomial
$16 \cdot y^5 - 16 \cdot y^3 + 3 \cdot y = 0$,
where $y = \cos(x)$. A factor of y can be removed, and the resulting equation is a disguised quadratic in y^2.

However, a better way of solving this equation is to use the Arithmetic Progression trigonometric specialist, described in section 2.3.1.9 above.

3.1.7 Limitations and Possible Extensions

This section discusses some of the limitations of the current implementation.

1. The base class is restricted to algebraic equations. This is an easy restriction to lift:

---

[12] In fact, the question asks the student to show that $x = 30$ degrees is a root of the original equation. This acts as a hint to the student that $y = 1/2$ is a root of the cubic. PRESS cannot use such hints.

requiring only an alteration of the parser and the acceptance conditions of the rewrite rules. However, the equations that have been encountered that might require such a modification, are better solved using other methods, such as those described in appendix I.

2. Some of the special purpose techniques are restricted in that they cannot cope with all equations in their class, these restrictions are noted in the relevant sections above and in Appendix II. Most of these restrictions are problem driven, the current techniques are capable of solving most of the equations encountered. These restrictions could all be overcome by making more use of meta-level inference at this level.

3. At present Homogenization works in one step. The offenders set is found, the reduced term is chosen and the offenders set is rewritten if possible. In some equations, however, more than one application of Homogenization may be needed. Consider the following:

$$\sin(2^{x+1}) = \sin(2^x)$$

The offenders set consists of two sin terms, so the set is of type trigonometric. However, the g.c.d. of the angles involved cannot be calculated, due to the assumption that the angles are linear polynomials. Hence the selection methods for trigonometric sets fail, and the simplicity method (see section 3.1.3.3) is used. However, the rewrite rules cannot express

$\sin(2^{x+1})$ in terms of $\sin(2^x)$,

so PRESS fails to solve the problem. If Homogenization was called on the arguments, however, the problem would be recognized as one of the form $\sin(2 \cdot y) = \sin(y)$, where $y = 2^x$. PRESS can easily cope with the problem in this form. So the ability to call Homogenization on the arguments of the offenders set can be useful. On the other hand no problems encountered on the papers require such an ability. Again, better use of meta-level inference might overcome this problem.

3.1.8 Conclusions for Standard Homogenization

This section has described the Standard Homogenization method, which prepares equations for solution by Change of Unknown.

As with the other methods of PRESS, Homogenization has its own set of rewrite rules. Only these rules need to be considered when rewriting terms in the offenders set. These rules also ensure that only algebraic rewrites are used. Many of the rules used by Homogenization are extremely explosive. Some can loop, and make the equation more complex, e.g.

$$U^V \Longrightarrow (U^W)^{(V/W)}$$

$$\log_X Y \Longrightarrow 1/\log_Y X.$$

A system cannot allow these rules to be applied in an unrestricted manner, but all of these rules are needed in certain situations.

3.1.8.1 The Need for Specialists

The problem of selecting the reduced term is one that requires something more than a general purpose method. Although the simplicity method is often successful there are many problems for which it is inadequate (see section 3.1.5.1 above). Thus special purpose methods have been implemented for some classes of equations, and these have proved to be of great value.

3.2 Extended Homogenization

So far this book has concentrated on techniques for solving single equations in one unknown. However, the test set of problems contains many simultaneous equations. An example of such a problem is:

$\cosh(x) - 3 \cdot \sinh(y) = 0$

$2 \cdot \sinh(x) + 6 \cdot \cosh(y) = 5$ \hfill (A.E.B. 1973)

Originally PRESS had a method that allows it to solve some simultaneous equations, but it was found that some of the equations encountered required methods that were more sophisticated. It was discovered that an extension of Homogenization was sufficient to allow many of these problems, including the above example, to be solved. The process of Elimination, a technique commonly used in solving sets of simultaneous equations, was also examined.

The next sections describes the method originally used by PRESS, which is called the **Elementary Method**. Extended Homogenization and Elimination are described in the following sections.

3.2.1 The Elementary Method

This section describes the method originally implemented in PRESS, the Elementary Method.[13] It was devised by Alan Bundy.

Given a set of equations, $\{e_1, \ldots, e_n\}$, and a set of unknowns $\{x_1, \ldots, x_m\}$, the task is to find those values of the x_i which satisfy the e_i.

At any point there are three lists, X, E, and S. Initially, X is the complete set of x_i, in the order they appeared in the problem, and E is the e_i, similarly ordered. S is initially empty. The method falls into two parts, reduction and back-substitution. Reduction consists of the following steps.

1. If X contains only one element exit and proceed to back-substitution. Otherwise, consider the first element of X, x_j say. Delete this from X.

2. Find the first member of E which contains x_j. Call this equation the **x_j-equation**.

3. Try to solve the x_j-equation for x_j. If the attempt to solve this equation fails, backtrack to the previous step and try to find another one. If there are no more possible x_j-equations, exit with failure. Otherwise, an expression for x_j in terms of some of the other unknowns is obtained, the **x_j-expression**. Put this expression in S. Delete the x_j-equation from E.

4. Using the x_j-expression, substitute for x_j in the equations remaining in E.

5. Now recursively apply this procedure.

Each application of the above procedure reduces the number of unknowns in X by one, so it is guaranteed to terminate. Every time an unknown is deleted from X, it is substituted for in E, so the number of unknowns in the equations in E also reduces. A stage will be reached where there is just one unknown, x_k say, in X, and there will be an equation in E which contains x_k as the only

---

[13]This method is similar to that called **stripping** in [10].

unknown. (There may be more than one such equation.) This equation is solved for x_k.

Now, **back-substitution** is applied. This consists of substituting the value obtained for x_k in all the equations in S. After simplification, this will produce a value for another unknown, which is then substituted into the rest of the equations. This process of back-substitution is repeated until values have been found for all the unknowns.

Consider an example:
$$3 \cdot x + y = 5 \qquad \text{(i)}$$

$$x^2 + 2 \cdot y^2 - 3 \cdot x + 2 \cdot y + 2 = 0. \qquad \text{(Oxford 1976)} \qquad \text{(ii)}$$

From (i) the x-expression
$$x = (5 - y)/3 \qquad \text{(iii)}$$
is obtained by reduction. Using (iii) to substitute for x into (ii) produces:
$$(19 \cdot y^2 + 17 \cdot y - 2)/9 = 0 \ .$$
Solving this for y gives $y = -1$ or $y = (2/19)$. Back-substituting these values for y in (iii) gives $x = 2$ or $x = (31/19)$. Therefore, the solution of the set of equations is:
$x = 2$ and $y = -1$, or

$x = (31/19)$ and $y = (2/19)$.

The performance of this method can be improved by some heuristics, and this has been done for the current PRESS implementation. The first "heuristic" used is to ensure that the equation chosen contains the unknown that the system is trying to solve for. If there are alternatives the program could simply choose one arbitrarily, and backtrack to make a different choice later if necessary. However, there are other factors that should be used to decide which equation to choose.

For example, it may be the case that there is already an equation which contains only one unknown. If so, this equation should be used first to solve for first the unknown, and then the result should be substituted for the unknown throughout the set of equations. This may then give another equation with only one unknown so the process can be repeated. When the process no longer continue, the current Elementary Method can be applied.

If there is a choice of equations to solve for a particular variable, the 'easiest' should be selected. A possible heuristic for this would be to choose the equation with the fewest occurrences of that variable. However, there are other considerations as well. It is often the case that one equation is linear in one unknown and quadratic in the other. Such equations should always be solved for the linear unknown, to avoid introducing a disjunction at an early stage. For example, the equation
$$3 \cdot x + 4 \cdot y^2 = 19$$
should be solved for x rather than y. This yields the equation
$$x = (19 - 4 \cdot y^2)/3.$$
If y is solved for first, this produces the equations
$$y = ((19 - 3 \cdot x)^{(1/2)})/4$$

or

$$y = - ((19 - 3 \cdot x)^{(1/2)})/4,$$
and each term of this disjunction must be solved separately.

The author has implemented the linear heuristic in the current version of PRESS.

Scope and Limitations

The solution of simultaneous equations is a major concern of numerical analysis. Many methods have been devised for solving systems of linear equations, with numerical coefficients, see [24] for example. By comparison, on such systems the Elementary Method is extremely inefficient. However, the Elementary Method is applicable to a much wider class of problems, it is not restricted either to linear equations, or to numerical coefficients.

In theory, the Elementary Method is as good as any other, in that no solutions will be missed using it that would be found by other methods. In practice, the method often becomes overburdened with complex terms, and fails to find a solution in a reasonable time. For example, consider the following question:

$$\cos(x) + \cos(y) = 1 \qquad \text{(iv)}$$

$$\sec(x) + \sec(y) = 4 \qquad \text{(A.E.B. 1976)} \qquad \text{(v)}$$

Solving (iv) for x produces as one set of solutions[14]

$$x = n \cdot 360 + \cos^{-1}(1 - \cos(y)).$$

Substituting this value in (v) produces

$$\sec(n \cdot 360 + \cos^{-1}(1 - \cos(y))) + \sec(y) = 4.$$

This equation cannot be solved by PRESS, even if the principal value of x (i.e. set n to 0) is used, as simplification rules of the type[2]

$$\sec(\cos^{-1}(y)) \implies 1/y$$

have not been provided.

However, even if PRESS did have access to the simplification rules required, it would still require a lot of work to solve the resulting equation. It is clear that far too much effort is used. In particular, there is no need to solve (iv) for x, only for cos(x). Then using the fact that sec(x) is 1/cos(x) the equations can be quite easily solved. Presumably, this method is the one that the examiner expected to be used.

When is the Elementary Method is not the best approach? What is a better method? Extended Homogenization appears to provide the answer to the second question.

3.2.2 Application of Homogenization to Simultaneous Equations

This section describes the application of Homogenization to simultaneous equations. This application is called **Extended Homogenization**, to distinguish it from Standard Homogenization.

The method is best illustrated by an example. Consider the equations discussed above:

$$\cos(x) + \cos(y) = 1$$

---

[14] n is an arbitrary integer. The other set of solutions is

$$x = n \cdot 360 - \cos^{-1}(1 - \cos(y)).$$

[2] There are a large number of rules of this type even if the hyperbolic cases are neglected. It is preferable for PRESS to manage without them.

$\sec(x) + \sec(y) = 4$.

The Standard Homogenization parser is used. Parsing the equations with x as the unknown gives the **x-offenders set** $\{\cos(x), \sec(x)\}$, and parsing with y as the unknown gives the y-offenders set, $\{\cos(y), \sec(y)\}$. A reduced term from the x-offenders set is chosen as in Standard Homogenization. In this case the **x-reduced term** is $\cos(x)$, the y-reduced-term is $\cos(y)$. (Note that in general the x_i-offenders sets need not contain the same functions.) The terms in the offenders sets are rewritten as algebraic functions of the reduced terms, using the rewrite rules

$\sec(x) \Longrightarrow 1/\cos(x)$

$\sec(y) \Longrightarrow 1/\cos(y)$.

Substituting these rewrites into the equations, and replacing $\cos(x)$ by x1, and $\cos(y)$ by x2 produces

$x1 + x2 = 1$

$1/x1 + 1/x2 = 4$.

Solving these equations by the Elementary Method yields $x1 = 1/2$ and $x2 = 1/2$. Finally, solving $\cos(x) = 1/2$ and $\cos(y) = 1/2$ gives the solution to the problem.

Now this must be generalized. Firstly, a new concept must be defined, that of Homogenization **with respect to** a variable. In Standard Homogenization this variable is the unknown. In Extended Homogenization, to homogenize an equation Eqn with respect to x_i means to apply the process of Standard Homogenization to Eqn treating x_i as the sole unknown. If an expression is being homogenized with respect to x_j, a term which does not contain x_i is effectively a constant, and will not appear in the offenders set created by this operation.

The process of Extended Homogenization will now be described.

There is the list X which consists of the unknowns, the x_i, and the list E of the equations, the e_i. The basic process is to homogenize the equations with respect to each variable. This is done as follows:

- Consider the first element of X, x_j say. The equations in E are homogenized with respect to x_j.

- To do this, the set of equations in E is treated as one entity, a conjunction of equations. This is parsed to obtain the **x_j-offenders set**. This set is the union of the offenders sets produced by parsing each of the e_i separately with respect to x_j. (As noted above the terms which do not contain x_j are treated as constants during this parse, and thus are not put into the offenders set.)

- Using the Standard Homogenization process each member of the x_j-offenders set is rewritten as an algebraic function of the **x_j-reduced term**, which is chosen in the standard way. These rewrites are then substituted into the conjunction of the equations of E.

- The equations are now homogeneous with respect to x_j.

- The process is repeated with the next member of X, and continues until every member of X has been used. At this point the equations are homogeneous in every unknown.

- Now Change of Unknown can be performed. y_i is substituted for the x_i-reduced term, for every x_i in X. These substitutions are recorded in R, the **reduced term** list. R therefore

consists of equations of the form:
$$y_j = f_j(x_j)$$
where f_j is some function.

- The equations in E are now algebraic in the y_i, and they can be solved using the Elementary Method. This gives the values of the y_i, and substituting these values into R gives a set of independent equations for the x_i.

In some cases the equations may already be homogenized (or algebraic) with respect to some of the variables. This is detected by the x_i-offenders set being a singleton. If this singleton is x_i this variable need no longer be considered for Homogenization, so the algorithm proceeds with the next variable. If the singleton is some other term, $f(x_i)$ say, the substitution $y_i = f(x_i)$ is made, and the algorithm proceeds with the next variable.

If Homogenization succeeds, then the solution to the equations is obtained more neatly than by using the Elementary Method on its own. If all the x_i-offenders sets produced contain only the x_i the equations are algebraic, and the Elementary Method can be applied without attempting Homogenization. In other cases however, there seems to be no easy way of telling whether Homogenization should be attempted. Consider the following question:

$$\log_y(x) = 2 \tag{vi}$$

$$\log_2(x) + \log_2(y) = 3. \quad \text{(A.E.B. 1976)} \tag{vii}$$

In this case, (vi) can be solved for x to obtain $x = y^2$. Substituting this value in (vii) gives an equation for y which PRESS solves easily. Thus the Elementary Method is quite appropriate for this question. Homogenization fails if it is attempted, PRESS cannot predict this in advance.

To overcome this problem, the present implementation adopts the following strategy: If the equations are all algebraic the Elementary Method is used. Otherwise Homogenization is attempted. If Homogenization fails the Elementary method is tried. These tactics allow PRESS to solve examples such as the one above. However, time has been wasted trying to homogenize.

3.2.3 Elimination

Extended Homogenization increases the range of problems which PRESS is able to solve. However, Homogenization is not taught to A level students, although they may 'rediscover' the method while working on a particular example. A method they *are* taught is **Elimination**. This method will now be discussed, and compared with Extended Homogenization.

The term **Elimination** is often applied to solving systems of simple linear equations. The Elementary Method is fairly satisfactory for equations of this type, but Elimination is in fact applicable to a much wider range of problems.

The process of Elimination consists of transforming some of the equations in some way, and combining these transformed equations so that a variable is eliminated from the set of equations.

The explanation of Elimination begins with an example of linear equations.

3.2.3.1 The Method Of Elimination

Consider the set of equations

$3 \cdot x + 2 \cdot y = 9$ (i)

$2 \cdot x - 5 \cdot y = -13.$ (ii)

It is decided to eliminate x, say. To do this, multiply (i) by 2, and (ii) by 3, and subtract the resulting equations. A single equation in y remains, $19 \cdot y = 57$.

Hence y equals 3 and thus the value of x can be found from (i) or (ii).

Obviously, on such a simple example nothing has been gained. Consider a more interesting case.

$\cos(x) - 3 \cdot \sin(y) = 0$ (iii)

$6 \cdot \cos(y) = 5 - \sin(x)$ (iv)

Here, the equation
$\cos(x) = 3 \cdot \sin(y)$ (v)
is obtained from (iii), and
$\sin(x) = 5 - 6 \cdot \cos(y),$ (vi)
is obtained from (iv).

Now square (v) and (vi) and add the two resulting equations to obtain:

$1 = 9 \cdot \sin^2(y) + 25 - 60 \cdot \cos(y) + 36 \cdot \cos^2(y).$ (vii)

x has been eliminated from the equations. (vii) can be solved to give the value of y, and substituting this value in (iii) yields an equation for the value of x.

Comparison with Extended Homogenization

How does Elimination compare with Extended Homogenization? Using Homogenization on equations (iii) and (iv) above produces the offenders sets $\{\cos(x), \sin(x)\}$ and $\{\cos(y), \sin(y)\}$. Letting $x1 = \cos(x)$ and $y1 = \cos(y)$ produces the equations

$x1 - 3 \cdot (1 - y1^2)^{1/2} = 0$

$(1 - x1^2)^{1/2} + 6 \cdot y1 = 5.$

Solving the first equation for x1, and substituting the value obtained into the second gives:

$(9 \cdot y1^2 - 8)^{(1/2)} = 5 - 6 \cdot y1.$ (viii)

Squaring (viii) and simplifying gives

$27 \cdot y^2 - 60 \cdot y + 33 = 0,$

a quadratic in y1. This equation is also generated when PRESS solves (vii), obtained by applying Standard Homogenization to (vii).

Comparing the two methods of solution seems to indicate that Elimination offers no particular advantage over Extended Homogenization on this kind of problem. Both involve a squaring operation and the application of Homogenization, the relative order varying. Both arrive at the same quadratic. Thus given an implementation of Extended Homogenization it seems that Elimination is superfluous.

However, as Elimination is such a well known method this section discusses how it could be implemented.

Planning the elimination

When is Elimination possible, and how does it proceed if it is? The case of linear equations will not be discussed. In this case Elimination is always possible[3] but usually it is unnecessary.

At the time of writing Elimination has not been implemented. Therefore the following should be viewed as a possible method of implementation, rather than a working program.

For Elimination to succeed the equations must be of the form:
$$a_i \cdot f_i(x) + g_i(y,z,u,v, \ldots) = 0$$
where x is distinguished. Note that the a_i are constants, any of the other variables have been merged into the g_i.

To eliminate x a rule is needed which relates the f_i, or a set of such rules. The type of rules required are those where the RHS contains fewer occurrences of the unknown than LHS. Such rules can be viewed as modified Collection rules (see section 2.3.1.2), but in practice the rules used may not exist as Collection rules.

For Elimination, such a rule is rearranged to the form of LHS1 $\Longrightarrow$ RHS1 where RHS1 is free of the unknown. This is always possible, because in the simplest case the rule can be rewritten as LHS - RHS $\Longrightarrow$ 0, which is of the required form. We call rules in this form **Elimination** rules.

In the above example the rule used was
$$\cos^2(x) + \sin^2(x) = 1.$$

A method is required for determining if a suitable rule exists. It seems that Homogenization should be applicable. A possible approach could be the following: There is a set of equations $\{e_i\}$, and a set of unknowns, the $\{x_i\}$.

- Using the Homogenization parser, all the equations are parsed with respect to all the unknowns, as in Extended Homogenization. However, there is one important difference. Each equation produces an offenders set for each unknown. In Extended Homogenization, the union of all the x_i offenders sets is taken, to produce *the* x_i-offenders set. In Elimination this union is not performed, the offenders sets are kept separately. Parsing e_j with respect to x_i produces an offenders set which will be called $O_{i,j}$.

- The next step is to try to find an i such that for all j, $O_{i,j}$ is a singleton set. If there is such an i, this means that every occurrence of x_i is isolatable within the equation in which it occurs. Therefore, it is reasonable to try to eliminate x_i.

- The union over k of the $O_{i,k}$ is formed to produce the x_i-offenders set. From this set, a reduced term is chosen, $f(x_i)$, using the Standard Homogenization method.

- The next step is to try to rewrite each member of the x_i-offenders set in terms of $f(x_i)$. This is of course a step that occurs in homogenizing the equations with $f(x_i)$ as a reduced term.

---

[3] unless each unknown appears in only one equation, and each equation contains only one unknown, i.e. the equations are of the form $x_i + a_i = b_i$, where a_i and b_i are constants. Such a set of equations can be solved trivially.

HOMOGENIZATION

- If the above step succeeds, each occurrence of x_i can be eliminated using equation e_j, which contains $f(x_j)$. The rewrite rule is simply transformed into a Elimination rule, as above.

Consider this process on the example above.
$$\cos(x) - 3 \cdot \sin(y) = 0$$

$$\sin(x) + 6 \cdot \cos(y) = 5$$

The offenders sets produced are: $\{\cos(x)\}$, $\{\sin(x)\}$, $\{\sin(y)\}$ and $\{\cos(y)\}$. All four sets are singleton so any one is chosen, say $\{\cos(x)\}$. The x-offenders set is $\{\cos(x),\sin(x)\}$. Now attempt to rewrite each member of this set in terms of $\cos(x)$. The rewrite of $\cos(x)$ is trivial, and
$$\sin(x) => (1 - \cos^2(x))^{1/2}.$$
Transforming this to an Elimination rule produces:
$$\sin(x) - (1 - \cos^2(x))^{1/2} = 0 \qquad \text{(ix)}$$
The equations are transformed so that the term containing x is isolated. This gives:
$$\cos(x) = 3 \cdot \sin(y) \qquad \text{(x)}$$

$$\sin(x) = 5 - 6 \cdot \cos(y). \qquad \text{(xi)}$$
(ix) is now used to eliminate x. The left hand side of (ix) contains two subterms. The first, $\sin(x)$, occurs in (xi). (x) is used to produce the second subterm, applying transforms to both sides of the equation. Hence
$$(1 - \cos^2(x))^{1/2} = (1 - (3 \cdot \sin(y))^2)^{1/2} \qquad \text{(xii)}$$
Now (xii) is subtracted from (xi). The LHS of the equation produced matches the LHS of (ix). Hence the RH sides can be equated. This gives
$$0 = 5 - 6 \cdot \cos(y) - (1 - (3 \cdot \sin(y))^2)^{1/2},$$
and x has been eliminated.

The above method is not as neat as might be desired. For example, it may seem preferable to use the rule
$$\cos^2(x) + \sin^2(x) = 1$$
is preferable rather than rule (ix).

However, if this rule was used, equation (xi), would have to be transformed, which is not necessary in the above method.

3.2.4 Conclusions for Extended Homogenization

Various methods of solving simultaneous equations have been discussed. Some of these techniques have been implemented in PRESS, and an outline of how Elimination could be implemented has been given.

It was fairly easy to implement Extended Homogenization in PRESS. Much of the difficult work had already been done for Standard Homogenization, which is used by Extended Homogenization in PRESS.

It has been shown that Extended Homogenization seems to offer at least as much as Elimination for non-linear equations, and for linear equations the Elementary Method is adequate.

Homogenization can also be used to implement Elimination. As is often the case in the algebra domain, the principle problem is to constrain the search space. Homogenization requires very little search, and is thus a promising candidate for use in solving simultaneous equations.

3.3 Other Uses of Homogenization

So far, Homogenization has been described as an equation solving method. However, other uses have been found for Homogenization, and one is described in this section.

From a generalized viewpoint, Homogenization consists of three inter-related parts:

1. A parser which finds terms which "offend" in some way.

2. A way of choosing the reduced term.

3. A set of rewrite rules which express the offending terms as a relation of a reduced term.

In the case of equation solving, the offending terms are those which are not algebraic. The relation in 3 must be such that the new expression will not be considered as offending by the parser, after suitable changes, such as Change of Unknown in the case of equation solving. The rewrite rules must take into account which terms may be selected as the reduced term.

This description lifts Homogenization from the domain of equation solving, but where else can it be applied?

3.3.1 Homogenization in Unification

One such place seems to be unification with equality in an equational theory, see [32] for example.

In this domain, a theorem prover may need to unify terms that do not directly unify. The underlying equational theory can be used to rewrite the terms, and transform them into terms that can be unified. In general, the problem of unifying terms using rewrites is undecidable, so some constraints need to be applied in a practical system.

Consider an example. Suppose that the three terms
$X \cdot \cos(Y) \cdot \sin(a)$ (xiii)

$\sin(X \cdot Z)$ (xiv)

and

$(1 - \cos(U) \cdot \cos(V))^{1/2}$, (xv)

are to be unified.

The terms do not directly unify, so they must be rewritten. The first step involves finding the first place where the terms fail to unify. To do this, the dominating functors of the terms are compared. If the terms have the same dominating functor the arguments of the functor are then compared. In the example, the dominating functor of (xiii) is "$\cdot$", of (xiv) it is "sin", and the dominating functor of (xv) is "$\char`\^$" (the exponentiation functor). The terms do not have a common dominating functor, so equalities in the equational theory that relate them must be found. Suppose that the equation theory contains the equalities
$\sin(2 \cdot A) = 2 \cdot \cos(A) \cdot \sin(A)$

$\sin(B) = (1 - \cos^2(B))^{1/2}$.

The first equality relates the dominating functors of (xiii) and (xiv), and the second equality relates (xiv) and (xv). As a constraint, it is required that one of the terms in each equality *directly* unifies with the corresponding term. In fact, both (xiii) and (xiv) directly unify with the corresponding sides of the first equality. The unification produces the substitution

$\{A/a, X/2, Y/a, Z/a\}$.

(xiv) and (xv) still need to be unified. Taking into account the substitution, this means that the following need to be unified:

$\sin(2 \cdot a)$ (xvi)

and

$(1 - \cos(U) \cdot \cos(V))^{1/2}$. (xvii)

The second equality is used. As before it is required that at least one direct unification can take place. The term (xvi) unifies with the left hand side of the second equality, producing the substitution

$\{B/2 \cdot x\}$.

However, (xvii) does not unify with the right hand side, i.e.

$(1 - \cos(U) \cdot \cos(V))^{1/2}$

does not unify with

$(1 - \cos^2(2 \cdot a))^{1/2}$,

so more rewriting is needed. The terms fail to unify when the arguments of the "-" functor are compared. The two terms have different subterms in this position, namely

$\cos(U) \cdot \cos(V)$ and $\cos^2(2 \cdot a)$.

The two subterms have different dominating functors, "·" and "^". If the equational theory contains the equality

$T \cdot T = T^2$,

this can be used to unify the terms, producing the substitution

$\{T/\cos(2 \cdot a), U/2 \cdot a, V/2 \cdot a\}$.

The three original terms have now been unified, (xiii), (xiv) and (xv) as required. The final substitutions are:

$\{X/2, Y/a, Z/a, U/2 \cdot a, V/2 \cdot a\}$.

How can this process be viewed as an application of Homogenization? The part of the parser is played by the parser in the standard unification algorithm, which detects when terms do not unify, and finds the first point of disagreement. The equalities in the equational theory are the rewrite rules. The reduced term is the term that directly unifies with one side of the inequality. This differs from the reduced term in Standard Homogenization in that a reduced term is chosen for every pair of rewrites. Also, as the equalities in an equational theory can be used both ways round, it does not matter so much which term is used as the reduced term if there is a choice.

3.3.2 Generalization

Change of Unknown is very similar to the method of Generalization in the program-property theorem-prover of [6]. Before proving a theorem by induction, the Boyer-Moore theorem-prover is able to generalize it by replacing several occurrences of the same subterm by a new Skolem constant, e.g. (rev a) in

(equal
    (append (rev a) (append b c))

```
        (append (append (rev a) b) c)
)
```
is replaced by d to produce
```
  (equal (append d (append b c)) (append (append d b) c))
```
which is then solved by induction.

Work by Boyer, Moore and Aubin, [3], has concentrated on the question of when Generalization is to be done, which occurrences of a subterm are to be replaced and what additional assumptions may need to be introduced to prevent over-generalization. These are not issues in equation solving. There is never a danger of over-generalization and all occurrences of a subterm should always be generalized (i.e. changed to a new unknown).

The method of Homogenization is complementary to the work done by Boyer, Moore and Aubin, because it suggests how an expression may be prepared for Generalization: subterms which were not previously identical may be made so, in order to allow Generalization to proceed.

3.4 Conclusions

This chapter has described Standard Homogenization and its extension to simultaneous equations, Extended Homogenization.

Homogenization appears to be a powerful technique, with a large domain of applicability. For example, it has possible applications in theorem proving.

3.4.1 Humans and Homogenization

Standard Homogenization is a powerful equation solving method. However, human equation solvers are not taught the concept of Homogenization, at least not to our knowledge.

Instead, each type of equation is handled in a separate way. For example, the problem of solving
$$e^{3 \cdot x} - 2 \cdot e^{x} - 3 \cdot e^{-x} = 0$$
is not considered similar to the problem of solving
$$7 \cdot \sin^2(x) - 5 \cdot \sin(x) + \cos^2(x) = 0.$$
As has been shown, both equations can be solved by rewriting certain terms as algebraic functions of another term, with the aim of allowing the Change of Unknown method to be used.

What is gained by recognizing that apparently diverse solution methods share something in common? One advantage is that the steps involved can be specified at a fairly high level, and can thus be seen to be applicable to many other types of equations as well. Extended Homogenization developed from Standard Homogenization in this way.

Also, less code has to be written as major steps are applicable to all the problem subtypes. In the case of Homogenization, all the different equation types use the same code for following steps:

- the parsing of the equation to find the offenders set,

- the rewriting of the offenders terms as functions of the reduced term, and

- the Change of Unknown step.

The difference between the problem subtypes occurs only in the manner in which the reduced term is chosen.

However, the considerations above relate to efficiency and conceptual clarity, PRESS would still perform if Homogenization was implemented separately for each type of equation.

This will be mentioned in relation to LP in a later section. As implemented, LP cannot learn the concept of Homogenization, but it is able to learn that certain rewrite rules should be used to allow Change of Unknown to apply.

Another advantage is that the generalized method can be applied to a totally new situation, using the default simplicity method for choosing the reduced term. All the program needs is the relevant rewrite rules. In fact Homogenization for hyperbolics was first implemented in this way, a few rewrite rules were given to PRESS and the simplicity method was used to choose the reduced term. By observing the behaviour of PRESS, the author was able to derive the special purpose techniques needed for this class of equations.

CHAPTER 4

PRECONDITION ANALYSIS AND LP

4.1 Introduction

This chapter describes the learning technique of **Precondition Analysis**. Precondition Analysis is introduced in the context of the learning program, LP.

4.1.1 Precondition Analysis

Precondition Analysis is principally used to learn strategies for problem-solving. The program learns from examples of correctly executed tasks.

A program using Precondition Analysis contains two parts, the learning cycle and the performance part. In the learning cycle, the program is given an example of a correctly executed task.[1] The example may contain several individual steps, each step being the application of an operator. The program first examines the example, to find out which operators were used in performing the task. This stage is called **Operator Identification**. During this phase, the program may discover that it does not possess the relevant operator, and the user is asked to provide the necessary information.

Once this phase is complete the program builds an explanation of the **strategic** reasons for each step of the task. The explanation is in terms of satisfying the preconditions of following steps. We call this phase the **Precondition Analysis phase**, this name gives rise to the name of the entire learning technique. (In earlier papers, such as [66], this stage was called Method Justification.)

From this explanation, it builds a plan that is used by the performance element. We call the plans **schemas**. The performance element executes this plan in a flexible way, using the explanation to guide it. The explanations are used to make sensible patches to the plan if the plan cannot be used directly.

Precondition Analysis relies on having control information explicitly represented, programs using meta-level inference have this property.

Precondition Analysis distinguishes between steps in the plan that are very important, and those that are not. These important steps are called **major steps**. Portions of the schema are devoted to trying to ensure that these major steps can be executed.

---

[1] In the context of LP, this is a worked example showing a way of solving an equation.

Various parts of the Precondition Analysis technique are similar to parts of the learning methods used by other researchers, particularly Mitchell. Chapter 5 highlights the similarities and differences.

The remainder of this chapter expands this brief description of Precondition Analysis.

4.1.1.1 LP

Like PRESS, LP solves equations. However, LP can learn new methods automatically. This contrasts with PRESS, where new methods are added by writing new code.

LP is not restricted to learning new *methods* in the PRESS sense. Using Precondition Analysis, it learns **strategic information**, in the form of a schema, an equation solving plan, which LP executes in a flexible way to solve new equations.

LP can also assimilate new rewrite rules.

Like PRESS, LP uses meta-level inference. Much of the learning task of LP consists of learning **control information**. The task is made easier by having this information represented explicitly as meta-level statements.

LP learns by examining worked examples. It is intended to be able to learn a new equation solving technique from a single worked example. As Neves points out, [58], many textbooks give only one worked example before a set of test problems. This approach contrasts with that of the concept-learning programs described in [20]. These programs require several examples to learn a new concept.

However, usually humans do not learn fully from the worked example. Often humans will try the exercises with only a moderate understanding of the technique, and go back to study the example further if an exercise can not be solved. LP does not do this, it is very much a "one-pass" system. Instead, it attempts to gain as much as possible from the worked example, and, if it runs into difficulty, it uses its knowledge to try to overcome its problems.

LP has successfully learned many new equation solving methods and schemas. For example, it has learned how to solve the type of equations that PRESS solves with the arithmetic progression specialist (section 2.3.1.9), e.g.
$\cos(x) + 2 \cdot \cos(2 \cdot x) + \cos(3 \cdot x) = 0$ (A.E.B. 1972)
It has also been taught to solve equations that PRESS can not solve, such as
$\sin(2 \cdot x) + \sin(3 \cdot x) + \sin(5 \cdot x) = 0$ (A.E.B. 1973)

LP contains most of the PRESS methods to begin with, (see section 4.3.2.2), but it is possible to disable them, and LP is able to learn instances of most of these methods, see section 4.6.

4.1.2 The Organization of this Chapter

Throughout this chapter, LP is used to illustrate Precondition Analysis. However, LP contains several special features which may not be shared by programs using Precondition Analysis in other domains, and these possible differences are pointed out as they occur.

The next section provides a broad outline of LP. Section 4.3 describes basic concepts used by LP. Section 4.4 describes in detail how LP acquires new instances of these concepts. Section 4.5 describes how LP uses its new techniques to improve its equation-solving ability. Appendix V contains examples of output from LP.

4.2 An Outline of LP

LP divides into two sections. The first section is an equation solving program, this is the "performance" element, (see [72]). At any stage of the learning cycle, we judge the ability of LP by the type of equations it can solve. Chapter 2 described how PRESS solves equations. At the beginning of a learning session, LP will attempt to solve equations in the similar way to PRESS, but unlike PRESS, LP is also able to improve its performance by learning from worked examples. This learning process greatly alters the equation solving process.

LP has a limited ability to test the value of the plans that it learns. It contains a **problem generator** capable of generating new equations that are somewhat similar to those solved in the worked examples. It attempts to solve these new equations using its new schemas. This allows it to assess the value of the schemas that is has learned. The problem generator is described in appendix III.

4.2.1 Learning from Worked Examples - An Outline

4.2.1.1 Worked Examples

This section describes the characteristics of the worked examples used by LP.

A worked example shows the steps involved in the solution of an equation. The example shows various points in the solution process. It is arranged as a sequence of **lines**, each line is an equation, or a disjunction or conjunction of equations. Each line can be transformed into the next[2] by applying a sequence of legal algebraic operations. We call such a sequence a **step**.

A worked example should contain enough detail so that the reader (either a human or a program) can understand the technique being demonstrated, but there should not be so much detail that the important points are swamped.

Figure 4-1 shows one of the worked examples used by LP. The example shows the solution of the equation
$$\cos(x) + 2 \cdot \cos(2 \cdot x) + \cos(3 \cdot x) = 0. \hfill \text{(A.E.B. 1971)}$$
The angle measure is degrees, and n_1 and n_2 are arbitrary integers. We will use this example in our discussion.

This worked example is typical of those used by LP. Note that the examples contain no annotations or grouping information. Section 4.3.1 describes why this form of worked example is preferred to the alternatives.

4.2.1.2 Learning from Worked Examples

What can LP learn from a worked example?

Perhaps the easiest way to answer this question is to consider what human students can learn from the worked example. Suppose that a student was familiar with the all the PRESS methods described

---

[2]Sometimes this will not be the case if the worked example is arranged in a strictly linear fashion. See section 4.3.1 below.

$$\cos(x) + 2\cdot\cos(2\cdot x) + \cos(3\cdot x) = 0 \qquad \text{(i)}$$

$$2\cdot\cos(2\cdot x)\cdot\cos(x) + 2\cdot\cos(2\cdot x) = 0 \qquad \text{(ii)}$$

$$2\cdot\cos(2\cdot x)\cdot(\cos(x) + 1) = 0 \qquad \text{(iii)}$$

$$\cos(2\cdot x) = 0 \; \vee^3 \; \cos(x) + 1 = 0$$

$$\cos(2\cdot x) = 0$$

$$x = 180\cdot n_1 + 45$$

$$\cos(x) + 1 = 0$$

$$\cos(x) = -1$$

$$x = 180\cdot(2\cdot n_2 + 1)$$

Figure 4-1: A Worked Example

in this book, except for the Trigonometric Factorization methods (sections 2.3.1.9 and section I.3).

What happens as the student works through the example shown in figure 4-1?

At the lowest level, the student may not understand how one particular line follows from the previous one. In our example, the student will not understand how line (ii) follows from line (i).

The teacher may explain to the student that line (ii) follows from line (i) by adding the first and third cosine terms, using the identity

$$\cos(x) + \cos(y) = 2\cdot\cos((x + y)/2)\cdot\cos((x - y)/2). \qquad \text{(iv)}$$

This step is the only one which uses identities that the student does not know. The teacher has provided an identity, which is used from left to right as a rewrite rule. The student must learn under what circumstances this rule should be applied.

The student must check the remaining steps in the example. In fact, all the other steps use methods which are known to the student. At the end of this phase, the student knows which methods have been used to transform each line into the next (and that the first step used the new identity).

As well as the problem of not understanding how one line transformed into the next, the student may also have problems seeing the *point* of a step. In this case the student knows *how* the line is obtained from the previous one, but does not understand *why* the step is needed.

For instance, in the example, why were the cosine terms added? There are many answers to this type of question, including
      "It helps to solve the problem."
This answer is a last resort, reminiscent of the response of SHRDLU, [87], to a series of "Why did you do that?" questions:
      "Because you asked me to."

---

[3] $\vee$ denotes disjunction.

A more useful answer comes from examining the next step. This step is a Factorization Preparation step. It cannot occur unless the previous step has been performed. It then makes sense to ask why the Factorization Preparation step is important. The answer is that Factorization Preparation allows Factorization to apply. Factorization always reduces the complexity of the equation to two or more simpler equations. (In this case the two factors are so simple that Isolation can apply). This means that Factorization is always a desirable step. While students are rarely told this explicitly, they learn the importance of Factorization at an early stage, because it is used in so many examples. We assume that because the student implicitly recognizes the importance of Factorization, there is no great need to ask why the Factorization step is important, beyond the answer that it helps solve the problem. (Factorization is one of the key methods of LP, see section 4.3.2.3).

When trying to find the point of a step, it is easier to examine the worked example from the end. This is because the motivation for the final steps is usually clear, these steps actually solve the equation. The remainder of the steps are more manipulative, steps that allow the final methods to apply. In the example, the equation is eventually solved by Factorization, followed by Isolation. To enable the Factorization step, a Factorization Preparation step is needed. This step cannot occur unless the previous step takes place. This is the "point" of adding the two cosine terms in line (i) to obtain line (ii).

The Factorization method has a precondition that the equation is of the form
$$e_1 \cdot \ldots \cdot e_n = 0.$$
This precondition is satisfied at line (iii) in figure 4-1, but not at line (ii). Thus one of the effects of the step from line (ii) to line (iii) is to satisfy this precondition. Similar reasoning applies to the step from line (i) to line (ii).

Using this kind of reasoning, the student can find the "point" of every step in the example.

It is not enough for the student just to analyze the example. The example illustrates a certain general technique (or more than one) and the student must be able to extract this technique so that it can be used on other problems. For instance, the general point illustrated in our worked example it is sometimes a good idea to add trigonometric terms, as this will eventually allow the application of Factorization.

One problem here is to make "sometimes" more precise, i.e. under what conditions should trigonometric terms be added. A deep analysis shows that the technique shown in the example works if the right-hand side is 0, the left-hand side is the sum of three cosine terms with angles in arithmetic progression, and the terms with the largest and smallest angles have the same multiplicative coefficients, i.e. the equation must be of the form
$$a \cdot \cos(p) + b \cdot \cos(q) + a \cdot \cos(r) = 0,$$

where $q = (p + r)/2$. The student probably will not discover all these conditions on first examination. The student may use the technique on a number of problems, before finding an equation where not all these conditions are satisfied. The student may learn more of the conditions from the failure that occurs in such cases.

How does LP compare with the above account?

Like the student, at the lowest level, LP needs to learn new algebraic identities, which will be used as rewrite rules. The best way of discovering if the example uses new identities is to examine consecutive lines in the example from the beginning. Like PRESS, LP has a standard set of methods. At each stage, LP tries to find a method that transforms one line into the next. LP does not need to consider all its methods for each step. Meta-level inference is used to restrict the search. If it is successful, it notes which method has been used and proceeds to the next step. If no method can be found, this indicates that either LP does not know the identity, or that it has the identity stored as a rule in the rule set of a method that cannot be applied.

LP conjectures an instance of the identity and asks the user to provide the general rule. This process is described in sections 4.4.1.1 and 4.4.1.2.

LP also analyzes the example to find the "point" of the step. It works backwards at each step marking which preconditions of the following method are satisfied by the step. If the application of method M satisfies preconditions P of the following method N, we say that the **major aim** of the application of M was to satisfy P. This process is described in detail in section 4.4.2.

At the next level up from rewrite rules, LP needs to learn new methods. New methods are created to explain the application of new rules. Method creation is described in section 4.4.2.1.

The worked example is now divided into sections. This division is required for the equation solving process. For examples involving Factorization, the first section consists of the steps leading up to the application of Factorization, this splits the equation into a number of factors. If there are n such factors, the rest of the worked example is split into n sections, each section contains the steps involved in solving one factor. The sections can be recursively divided, e.g. in the case where the solution of one factor involves Factorization again. Each section has a **purpose**. The purpose of the first section of a Factorization example is to factorize the equation. The purpose of the remaining sections is simply to solve the factor. The concept of purpose is important when LP tries to solve new equations. Section 4.4.2.6 describes the division process in more detail.

A worked example shows how to solve one particular equation. However, students are expected to generalize the example so that they will be able to solve equations of a similar kind.[4]

LP stores its information about the worked example in a new method that contains a structure called a **schema**. This contains a list of methods that were used in the worked examples, tagged with its major aim and other information. The schema is divided into the same sections as the worked example. The schema encapsulates the lesson of the worked example, and is used to solve new equations. The schema is stored in a special kind of method, called a **schema method**

Like the student, LP does not always discover all the conditions relevant to the example. For example, it does not discover that the angles need to be in arithmetic progression in the worked example above. However, this does not matter very much because of the way that the schema is executed.

When LP attempts to solve equations, it uses the schema as a plan. By executing the plan in a flexible way, LP can solve equations that are related but fairly different from the original equation.

4.2.2 Solving Equations - An Outline

When LP is given a new equation, the **given equation**, to solve, it attempts to find a schema method that may help guide the solution process. A schema method may be appropriate if its preconditions are satisfied by the given equation. If there are several such schema methods, LP applies additional tests.

---

[4]They may also be able to obtain information from the worked example that will prove useful in solving equations that are quite different from the sample equation. For example, new rewrite rules may be applicable to many different types of equations.

If no suitable schema can be found, LP works in a similar way to PRESS, using the heuristic waterfall control method (see section 2.3.3).

Usually a suitable schema method will be found. LP now uses the information in the schema to help it solve the given equation. The schema lists the steps used to solve the generating equation.

A reasonable first strategy for LP is to attempt to apply exactly the same steps to the given equation. If the given equation is very similar to the generating equation, this attempt may be successful. In the general case, however, there will be significant differences between the two equations.

If this is the case, LP has to "patch" the schema. There are several ways of doing this. Firstly, LP may be able to omit steps. As mentioned above, the schema is divided into parts, each part having a purpose, e.g. factorizing the equation. At each stage, LP examines the equation to see if the purpose has already been achieved.[5] If this is the case, LP proceeds to the steps in the next division. This check sometimes enables LP to omit large parts of the schema. Usually, however, LP discovers that the purpose has not been achieved.

So, in general, LP will reach a point where it will not be able to apply the next step listed in the schema. Suppose that this step is an application of a method M.

At this point, LP uses the other information contained in the schema. Each step in the schema is tagged with its major aim, i.e. the set of preconditions the step is intended to satisfy. Suppose that the major aim of the application of M is to satisfy the set of preconditions ME.[6] LP searches for another method that is also capable of satisfying ME. If it finds one, M_1 say, LP attempts to apply M_1 to the equation. If this application is successful, LP continues with the next step in the schema, i.e. the step *after* the application of M. M_1 has been substituted for M. If the original worked example was analyzed correctly, and the only aim of the application of M is to satisfy ME, there should be no problem. Of course, this is not always the case. See section 4.4.2.5. If LP cannot find a method M_1, or M_1 cannot be applied, it looks for a method L that does not undo any already satisfied conditions, and attempts to apply it. If this succeeds, LP then tries to apply the method M. There is no good reason to suppose that M will now apply, this is a last resort to try to use the schema. If M cannot be applied, LP can repeat this process, until it cannot find a method L, or M can be applied. If M can be applied, LP continues as before with the next step in the schema.

If all of the above steps fail, LP may be able to backtrack. For example, there may be more than one choice for L above. If all such choices fail, LP can try to select another schema method, and use that. If all backtracking possibilities have been exhausted, LP goes back to the original equation, and behaves like PRESS, using a waterfall to guide it. After each transformation, it can try again to use a schema method.

4.3 Basic Structures of LP

This section describes in more detail four structures used by LP:

1. worked examples,

---

[5]PLANEX operates in a similar way, see section 5.3.

[6]Major Effects.

2. methods,

3. rewrite rules,

4. the Condition and Connection Tables.

Worked examples provide instances of new equation-solving techniques.

Methods and rewrite rules are used by LP to transform equations. In order to improve its equation solving performance, LP may need to learn or assimilate new instances of these two structures.

The Condition and Connection Tables describe the relations between the various methods and conditions. These following sections describe examples, methods, rewrite rules and Condition and Connection Tables in that order.

4.3.1 The Arrangement of Worked Examples

Section 4.2.1.1 described various characteristics of the worked examples used by LP. These examples are simply lists of equations. There are other ways of arranging worked examples. The ones used by LP differ from those in books in two ways.

Firstly, our examples contain no annotations. If a worked example comes from a textbook, it is likely that the example will contain annotations. These annotations give explanations where difficulties may arise, so that in the example of figure 4-2 (which is a reproduction of figure 4-1 above), between the lines (v) and (vi).

$$\cos(x) + 2\cdot\cos(2\cdot x) + \cos(3\cdot x) = 0 \qquad \text{(v)}$$

$$2\cdot\cos(2\cdot x)\cdot\cos(x) + 2\cdot\cos(2\cdot x) = 0 \qquad \text{(vi)}$$

$$2\cdot\cos(2\cdot x)\cdot(\cos(x) + 1) = 0 \qquad \text{(vii)}$$

$$\cos(2\cdot x) = 0 \ \vee^7 \ \cos(x) + 1 = 0 \qquad \text{(viii)}$$

$$\cos(2\cdot x) = 0 \qquad \text{(ix)}$$

$$x = 180\cdot n_1 + 45 \qquad \text{(x)}$$

$$\cos(x) + 1 = 0 \qquad \text{(xi)}$$

$$\cos(x) = -1$$

$$x = 180\cdot(2\cdot n_2 + 1)$$

Figure 4-2: A Worked Example (repeated)

$$\cos(x) + 2\cdot\cos(2\cdot x) + \cos(3\cdot x) = 0$$

$$2\cdot\cos(2\cdot x)\cdot\cos(x) + 2\cdot\cos(2\cdot x) = 0$$

---

[7] $\vee$ denotes disjunction.

there will be a phrase such as
> "Adding the first and third terms."

It may also give the identity

$\cos(x) + \cos(y) = 2 \cdot \cos((x + y)/2) \cdot \cos((x - y)/2)$.

LP does not use annotated worked examples, partly because of the difficulties involved with natural language processing, and also because we are interested to see how much can be learned without such aids. Neves, [58], adopts a similar approach with his ALEX program. (Section 5.2 describes ALEX.)

The second difference concerns the arrangement of the worked example. The worked examples used by LP are arranged in a strictly linear fashion. This can cause problems if the worked example contains Factorization or Change of Unknown steps. Consider figure 4-2 again.

Generally, we assume that each line in the worked example follows from the previous one, e.g. the step from (vii) to (viii) is recognized as a Factorization step. However, this assumption breaks down if the example is presented linearly in the way we adopt, line (ix) does not really follow from line (viii), and the line (xi) certainly does not follow from (x).

LP overcomes such problems by looking out for special situations, such as Factorization and Change of Unknown. In this case, when it reaches (viii), it expects to find the separate solutions for each of the factors. It looks ahead firstly for the step (ix) and then for the step (xi). If it fails to find them, a warning message is issued to the user. Then all steps between (ix) and (xi) are treated as solution steps for the equation (ix).[8] Similarly all steps after (xi) belong to equation (xi). Generally, if the Factorization step produces n non-trivial factors, LP will look for n separate solution sequences.

In the Change of Unknown Case, the corresponding situation is looking for the changed variable equation and the substitution equation and partitioning the worked example suitably. For instance, if the example is

$4 \cdot \log_x 2 + \log_2 x = 5$

$4/\log_2 x + \log_2 x = 5$

$y = \log_2 x$ (xii)

$4/y + y = 5$ (xiii)

$y = 1 \vee y = 4$ (xiv)

$\log_2 x = 1 \vee \log_2 x = 4$ (xv)

$\log_2 x = 1$

$x = 2$

$\log_2 x = 4$

$x = 16$

Line (xii) is the substitution. Equation (xiii) is the changed variable equation. Solving this gives the line (xiv). Substituting back gives the substitution equation (xv). Note that the substitution equation

---

[8] In this case of course there is only one step. In general, there may be an arbitrary number.

is a disjunction, and each disjunct is solved separately, as in the Factorization case. LP is also able to detect this.

The additional complexity is caused by the decision to present the worked examples in a linear way. Worked examples in books avoid this problem in one of two ways. Some books group the solutions of individual factors together, using indentation or boxes, perhaps showing the solution of several factors in parallel, each in its own box. The other solution uses annotations. Before each factor is solved, a comment to this effect appears in the text.

Despite the problems it causes, we prefer to use the linear form for the worked examples, because of its simplicity compared to the grouping or annotation approaches.

A similar problem may arise in other domains. For example, consider a repair task. The first part of the task may involve accessing one large component. Once this component has been accessed, it may contain several subparts that can be processed in any order. A trace of the task will show the one particular order that was used, in a similar way to the Factorization example.

In order for Precondition Analysis to be used in such cases, the program must know how subtasks may fit together, and be able to untangle the linear presentations.

4.3.2 LP Methods

LP methods are similar to PRESS methods. As with PRESS, each LP method has a set of preconditions and a set of associated rewrite rules. They also have a set of **postconditions**, facts which must be true after the method has been applied, see next section. Some methods also contain extra planning information, the schemas referred to above. This information is described in more detail in section 4.4.2.7.

4.3.2.1 Preconditions and STRIPS

Many learning programs work in domains where the basic operators are similar to those of STRIPS, [35, 36]. LEX, [55, 56], SAGE, [44], and ALEX, [58] are examples. The operators of STRIPS have preconditions, an **add list** which contains the facts that the operator makes true, and the **delete list** which contains the facts that are no longer true after the application of the operator. The facts on the add and delete lists are sometimes called the **effects** of the method, the method makes these facts true. If the preconditions are satisfied the operator can be applied, producing the effects. The preconditions of STRIPS operators are both sufficient and necessary.[9]

In contrast, LP methods do not have this desirable property. The preconditions of a method represent necessary, but not sufficient conditions. In general, a method is not certain to succeed, even if the preconditions are applicable. This is because the preconditions are too general, but we cannot give stronger preconditions that do not involve actually applying the method to test if it is applicable! It seems that this might be a problem in many domains.

The same comments apply to the preconditions of the PRESS methods, described in section 2.4. However, as PRESS does not plan or learn, the STRIPS-type property is not so important.

---

[9]Or rather, this is the usual state of affairs. Although STRIPS has some limited ability to cope with situations where the operator does not produce the intended effect, or cannot be applied, these are considered to be exceptional cases. See section 5.3 for further details.

Postconditions and Effects

Similarly, the effects of a method are hard to classify.

LP uses postconditions rather than effects. The postconditions are used to specify what must be true after a method has been applied in the **desired** way, but there is no guarantee that the method will make these true, i.e. it may be possible to apply the method in an undesirable way, making the postconditions false. Postconditions can be used to check that the method has been applied in the right way.

The fact that LP methods lack these properties is important. If the methods were as well-behaved as STRIPS operators, it would be possible to build a much more powerful planning system, and a lot of the heuristic nature of LP would be unnecessary. STRIPS is further discussed in section 5.3.

Despite these differences, LP methods play the role of operators in the general descriptions of Precondition Analysis.

4.3.2.2 LP's Initial Methods

The current version of LP initially contains the following methods:

1. Attraction

2. Change Of Unknown

3. Collection

4. Factorization

5. Factorization Preparation

6. Homogenization

7. Isolation

8. Logarithmic Method

9. Nasty Function Methods

10. Polynomial Methods

These methods are the same as the PRESS methods of the same name[10] except that, to keep the program to a reasonable size,[11] some of the less frequently used procedures and rules have been removed. In particular:

- All rules and procedures relating to hyperbolic functions have been removed from LP. This affects Isolation and Homogenization.

- The Inverse-Trigonometric Elimination methods (described in section I.2.4) have been

---

[10]The Logarithmic and Nasty Function Methods are described in appendix I, Homogenization is described in chapter 3, the other methods are described in chapter 2.

[11]The current version of LP contains 2460 Prolog clauses, defining 1000 predicates.

removed from the Nasty Function Methods.

- The symmetric and anti-symmetric code has been removed from the Polynomial Methods.

LP is not restricted to its initial methods, it is also able to create new ones. This process is discussed in section 4.4.2.1.

4.3.2.3 The Key Methods

Certain methods are considered **key methods**. These are methods of special importance. A key method is one that should be applied whenever possible, in preference to other methods. For LP, there are two classes of key methods. The first are those that lead directly to solutions, Isolation and Polynomial Methods. The second class consists of Factorization and Change of Unknown.

In fact, Precondition Analysis would work without the concept of key methods, but the classification is useful in two ways.

As described above in the outline section 4.2.1.2, the Precondition Analysis phase explains the purpose of each step in terms of the satisfaction of the preconditions of the following steps. However, key steps do not have to be justified in this way, they are always desirable. Thus, key methods can be used to terminate the explanation early, producing the explanation

This step was performed because it is always a good idea to do so!

The effect of this is to divide the example into sections, each section preparing for an application of a key method. We expect the sections to fit together in some coherent way by our choice of key methods.

Obviously, if the only key methods are those that always lead directly to a solution, we have gained nothing, the only section is the whole example. However, if there are other methods as well, the explanation process is simplified, as there will be several short sections instead of one large one. The set of shorter explanations is more understandable to the human user, provided that the program and user agree which of the methods are key. This process is discussed in more detail in section 4.4.2.

The division into sections also has advantages when solving new equations. Basically, the key method gives each section a purpose, to apply the key method. The program always tries to achieve this purpose as soon as possible, preferably by applying the key method next. This sometimes allows the program to omit several steps, shortening the solution process. Again, this process is more understandable to the user than a long sequence not using key methods, provided that there is agreement as to which methods are key.

What makes a method a key method?

Methods that lead directly to a solution should be key methods. To decide which of the other methods are key methods requires a good knowledge of the domain. Given enough examples, a program may be able to discover which methods are applied whenever possible, and thus be able to find all the key methods. As LP is intended to learn from one example, it does not have this ability. If the program is given only a few examples, it must be told which of the methods (which do not lead directly to a solution) are to be considered key, as there is no other way for it to find out.

Consider the choices made by the author for LP. In the case of Isolation and Polynomial Methods, these methods lead directly to a solution and so are obviously desirable.[12] What about the other two methods, Factorization and Change of Unknown?

Both of these key methods simplify the current equation, at the expense of creating more than one subgoal. Factorization splits a product into individual factors, each of which must be solved. Change of Unknown substitutes a new variable for the occurrences of a subterm in the equation, and produces two new problems, the changed variable and substitution equations. Each individual factor produced by Factorization is simpler than the original equation, because the factor contains only a subset of the symbols occurring in the original equation. Similarly, the changed variable equation is simpler than the original equation as a whole subterm has been replaced by a single variable. (Recall that the term which is substituted for cannot be the original unknown.)

However, a method does not become a key method just because it produces a simpler equation. If Collection applies, it will often produce an equation that contains only a proper subset of the symbols in the original equation, and it only produces a single equation. Collection is nevertheless not a key method, as it does not meet the requirement that it should always be applied. There are many examples where an application of Collection is an undesirable move.[13]

In summary, it is helpful to define various methods to be key as this makes both the learning and problem solving processes more understandable to the human user. However, Precondition Analysis will work even if no methods are defined as key.

4.3.3 Rewrite Rules

LP has sets of rewrite rules for Isolation, Collection etc, but a worked example may use a new algebraic identity. LP must be able to detect that that has happened, so that it can learn the identity, and use it as a rewrite rule.

The first task is not difficult; if LP cannot find another explanation for a step, it assumes that a new rewrite rule has been used. The step may have used an existing method, and LP simply lacks the needed rule, or the step may have been an application of a new method. (LP can distinguish between these two cases, see section 4.4 below.)

A worked example will only use a specific instance of a rewrite rule, and the program needs the general case. A concept-learning program could perhaps discover the rule after being shown several examples, but LP only has one example.

At present, LP finds the specific instance, and asks to be given the general rule. A more general system might be able to conjecture a possible rule, and attempt to prove it correct. However, we are interested in learning control information, and the learning of new rules is not our focus of attention. Therefore, we have not given LP the ability to conjecture general rules.

By finding the specific instance, LP helps the user find which general rule is required. LP also

---

[12]Note that other methods *can* produce solutions. For example, applying Collection to
$x + \cos(x) - \cos(x) = 0$
produces the solution $x = 0$. However, these are special cases, and unlike Isolation and Polynomial Methods, Collection can be successfully applied without producing a solution.

[13]For example, applying Collection to the first term on the LHS of
$\sin(x)\cdot\cos(x) + \sin(x) = 0$
(to get $(1/2)\cdot\sin(2\cdot x)$) prevents the application of Factorization Preparation.

checks that the general rule it is given does in fact account for the specific instance.

Section 4.4.1.2 describes how LP finds the specific instance.

4.3.4 The Condition and Connection Tables

During the equation solving process, LP may need to "patch" the schema, e.g. it may need to substitute one method for another, or add an extra method etc. In order to patch correctly, LP needs to be able to answer the questions:

- Will applying method M destroy the satisfaction of condition C?

- Will applying method M achieve condition C?

To answer these questions, LP maintains two tables.

The first table records the connection between various conditions, e.g. that an equation cannot satisfy both single-occ(X,Eqn) and mult-occ(X,Eqn). This table is called the **Table of Conditions**. As LP uses a fixed set of conditions for its preconditions and postconditions, this table can be created once and for all. For each condition C the table records up to two pieces of information:

1. Which conditions exclude C, and

2. which conditions imply C.

For quick access, the information is stored in a redundant manner, e.g. recording both that C excludes C', and that C' excludes C.

LP also contains a **Table of Connections**. This table records which methods *might* cause the satisfaction of various conditions, indexed off the **conditions**. For example, it records that the single occurrence condition may be satisfied by the application of Collection. Often, more than one method is capable of satisfying a condition, e.g. the common subterms condition may be satisfied by Collection, Attraction and Homogenization. This information as a whole is not available elsewhere, although part of it is. The postconditions of the methods indicate which conditions various methods *must* satisfy. However, this information does not show that, e.g. Collection can satisfy the single occurrence condition, as this is not a necessary condition for Collection. In fact, for ease of use, the table differentiates between the methods that must satisfy the condition, and those that might. As the latter information could be found from the postconditions, this table also contains redundant information. However, it is helpful to have it in the table, as the plan patching process needs to have this indexing off conditions.

Unlike the Table of Conditions, the Table of Connections is partially updated as new methods are created, see section 4.4.2.2.

Section 4.5.2 describes how these tables are used during the schema patching process.

4.4 How LP Learns

Section 4.2.1 above outlines how LP analyzes a worked example. This section describes the process in more detail. We will use the worked example shown in figure 4-2 above.

4.4.1 Operator Identification

As described in section 4.2.1, the first task for LP is to discover **how** each line in the worked example is transformed into the next. This process is called **Operator Identification**.[14]

Usually,[15] the program examines consecutive pairs of steps, i.e. it examines line one and line two, then line two and line three, etc. Note that for Operator Identification the example is examined in a forward direction, from the initial equation. This contrasts with the backward examination that takes place when LP attempts to find the aim of each step, see section 4.4.2.

Suppose that LP is working on the step from a line l to the next line, called m. LP first tries to see if an existing method can account for the step. To do this, LP tries to find a method whose preconditions are satisfied by l, and whose postconditions are satisfied by m.

If LP finds such a method, it attempts to use it to transform l into m. If the method is successful, LP records that the step from l to m was performed by that method, and proceeds to the step from m to the next line.

For example, consider the worked example shown in figure 4-2 above. Suppose that LP is working on the step from line (vi) to line (vii), i.e.

$2 \cdot \cos(2 \cdot x) \cdot \cos(x) + 2 \cdot \cos(2 \cdot x) = 0$

$2 \cdot \cos(2 \cdot x) \cdot (\cos(x) + 1) = 0$.

LP discovers that Factorization Preparation can transform line (vi) to line (vii).

In the interesting case, the program cannot find a method that would account for the step, and it marks the step as **not immediately parsed**.

Suppose that the program has a step that it cannot account for, i.e. no known method would produce the transform between lines l and m. There are two distinct cases.

The first case is when LP finds a method, M say, whose preconditions and postconditions are satisfied by the two lines, but it finds that M cannot produce the transformation. In this case it informs the user that it is probably missing a rewrite rule for the method M. At a later stage the user is asked to provide the missing rule, and this rule is then made available to M, see section 4.4.1.2.

For example, the number of occurrences of the unknown might be less in m than in l, indicating that Collection might be a possibility. However, the attempt to apply Collection to m fails. In this case the program conjectures that the step *is* done by that method, but it does not have the necessary rules. In some early runs of this program, it conjectured that a certain step seemed to be done by Collection, but it could not do it. It had not been expected that this would occur on the example being used, and, on checking a gap was discovered in the Collection method of PRESS!

The second case is when no apparently applicable method is discovered. The program then assumes that an unknown rewrite rule (i.e. an identity) has been used by a new method.

---

[14] Operator Identification was originally referred to as **Step Justification** in [67, 68].

[15] As discussed above in section 4.3.1, sometimes the program does not examine consecutive pairs. This occurs when the example contains Factorization or Change of Unknown.

In either case[16], the program can use the worked example to conjecture the particular instance of the rule.

4.4.1.1 Conjecturing an Instance of the Rule

To do this, LP makes use of a very obvious, but extremely effective, heuristic. The heuristic can be paraphrased

> "Given two consecutive lines of an example, delete all terms common to both, and conjecture that the remaining expressions are equal."

For instance, given
$$A + B = C$$
$$D + B = C,$$
as two consecutive lines of a worked example, it is reasonable to conjecture that $A = D$. It has to be decided what deleting all common terms means. Could the same conjecture have been obtained if the second line above had been
$$B + D = C$$
instead? The answer depends on what level of knowledge is being assumed. At the A level Mathematics standard it is known that addition is a commutative operator, so the two occurrences of B can be deleted. On the other hand, if the example was
$$A^B = C$$
$$B^D = C$$
the conclusion does not go through, as exponentiation is not commutative.

Consider the worked example shown above in figure 4-2. For example, suppose that the program does not immediately parse the step from (v) to (vi), reproduced below.
$$\cos(x) + 2 \cdot \cos(2 \cdot x) + \cos(3 \cdot x) = 0$$
$$2 \cdot \cos(2 \cdot x) \cdot \cos(x) + 2 \cdot \cos(2 \cdot x) = 0$$
Using the heuristic, the program notes that both lines contain the additive term $2 \cdot \cos(2 \cdot x)$. It deletes this and considers the rest. Nothing else is common, so it produces the correct conjecture
$$\cos(x) + \cos(3 \cdot x) = 2 \cdot \cos(2 \cdot x) \cdot \cos(x) \tag{i}$$

Note that the common terms deleted must be dominated by the same function symbol. Otherwise, the program might delete the $\cos(x)$ term that occurs in both sides of (i) above, and produce the *wrong* conjecture
$$\cos(3 \cdot x) = 2 \cdot \cos(2 \cdot x).$$

The heuristic is not perfect, LP can make errors. Given the two lines
$$\sin(x) + \cos(x) = 0$$
$$\tan(x) + 1 = 0,$$
LP will incorrectly conclude that
$$\sin(x) + \cos(x) => \tan(x) + 1, \tag{ii}$$

---

[16] There is actually a third possibility; the step may be an application of an unknown method, using a known rule. LP checks to see if one of its *new* rules, (i.e. one which it has been given earlier in the session) could account for the step. See section 4.4.2.4. However, this case is actually very similar to the second case, the program must construct a new method. The only difference is that the user does not have to provide the new rule in this third case, as s(he) has already supplied it.

is an instance of the rewrite rule
$$\sin(Y) + \cos(Y) \Longrightarrow \tan(Y) + 1. \tag{iii}$$
(Note that (ii) is an instance of (iii), (iii) is universally quantified over Y.) In this case the problem can be avoided by giving intermediate lines, e.g.
$$\sin(x) + \cos(x) = 0$$

$$\sin(x)/\cos(x) + \cos(x)/\cos(x) = 0/\cos(x)$$

$$\sin(x)/\cos(x) + 1 = 0$$

$$\tan(x) + 1 = 0,$$

but this is undesirable as we prefer to keep the number of lines in the worked example as small as possible. Instead, we have to allow LP to make an occasional mistake. LP asks the user if the conjecture is correct before proceeding with further processing, see section 4.4.1.2.

Neves uses a similar heuristic in his program ALEX, [58]. However, ALEX is *not* given the rule, it guesses the rule by generalizing from the example. This is done by replacing numbers with variables where this appears to be appropriate. However, ALEX can be misled by spurious correlations. For example, suppose the example contains the lines is
$$2 \cdot x + x = 3$$

$$3 \cdot x = 3.$$
ALEX may wrongly generalize this to
$$N \cdot x + x = A \Longrightarrow A \cdot x = A.$$
(SIMPLE, [26], a diagnostic modelling version of ALEX using the analogy-matching techniques of Evans, [34], produces (intentionally) similar mal-rules.)

LP does not have this problem as it does not generalize. It does not run into this problem when producing conjectures either, as it is conservative in its assumptions.

In general domains, it should be possible to conjecture the instantiation of the missing object knowledge. This is done in the same way as here, simply by considering what has changed between the two states, and using additional knowledge. In the example, we used the fact that addition is associative and commutative. The equivalent piece of knowledge in other domains may be knowing that the order of certain objects is immaterial.

4.4.1.2 Acquiring new rewrite rules

LP has conjectured a specific instance of the rewrite rule. It now asks the user to confirm the conjecture. If the conjecture is not true, the program exits with failure. In some cases, it may be possible to produce alternative conjectures, but LP does not do this, mainly for efficiency reasons.

If the conjecture is true, the user is asked to provide the general rule. For example, if the conjecture was (i) above, the user would provide the rule
$$\cos(A) + \cos(B) \Longrightarrow 2 \cdot \cos((A+B)/2) \cdot \cos((A-B)/2). \tag{iv}$$

As a check, the program then tries to apply the new rule to the LHS of its conjecture, and tries to obtain the RHS. If it fails, this usually means the rule has been mistyped and the user is asked to retype it.

4.4.1.3 Storing the New Rule

Now the rule must be assimilated. Recall that there are two possible reasons why a step may be not immediately parsed. Firstly, the step may use an existing method M^{17} but LP may lack the necessary rewrite rule, or the step may use an entirely new method.

In the first case LP adds the new rule to the rule set of M. Now method M should be able to perform the step. This process allows LP to add new rules to methods, thereby increasing their scope.

Otherwise, LP must create the new method, the new rule will eventually be placed in the rule set of this new method.

SCOPE

Davy's program SCOPE, (unpublished) also addressed the problem of assimilating new rules into a PRESS-type system. Davy's solution differs from the approach used by LP.

LP classifies the rule by examining the way that it was used in the example, (assuming that the rule was applied by a known method.) The input to SCOPE is an identity. SCOPE treats the identity as two rules, one from left to right, and the other from right to left. It considers every word (atom) in the identity in turn as the unknown. Thus a single identity may be classified in more than one way. Thus, the identity
$$A^B \cdot A^C = A^{B \cdot C}$$
is classified as

1. A Collection rule for A from left to right.

2. An Attraction rule for B and C from left to right.

LP could be extended to perform this kind of classification. However, most of the rules we supply to LP can only be classified in one way.

4.4.1.4 Proof Checking

Operator Identification is similar in some ways to **proof checking**, a branch of automated theorem-proving. Proof checkers are given partial proofs, proofs with many steps omitted, the kind of proof a human mathematician might produce. The job of the proof checker is to fill in the missing steps, i.e. to turn the partial proof into a complete proof. If this can be done it shows that the partial proof is indeed valid.

Operator Identification is similar to proof checking in that it has to discover how one line of the example follows from the previous one. Also, the conjecture of the instance of the rewrite rule is similar to the generation of a lemma in theorem-proving. If a proof checker cannot find a proof connecting two statements in the partial proof, it may nevertheless be able to construct a lemma. If the lemma is true, the program can complete the partial proof.[18] The program cannot prove the lemma on its own, and the user is asked to help.

Operator Identification differs from proof checking in that the proof checker may have to fill in several missing steps between consecutive lines in the proof. Operator Identification assumes that

---

[17] Note that M need not be one of the original methods, M can be a method created by LP earlier in the session.

[18] The lemma can be constructed in various ways. The program can work forward from one statement and backwards from the following one, producing the lemma when the two searches seem to be "close enough", a heuristic requirement.

there is (at most) only one missing operator between consecutive lines. This makes the search space in Operator Identification much smaller than that in proof checking.

4.4.2 The Precondition Analysis Phase

Figure 4-3 shows the example of figure 4-2 after Operator Identification. Each line is preceded by the name of the method that produced it, or an explanatory comment.

$$\cos(x) + 2\cdot\cos(2\cdot x) + \cos(3\cdot x) = 0 \qquad (v)$$

(Applying New Rule)

$$2\cdot\cos(2\cdot x)\cdot\cos(x) + 2\cdot\cos(2\cdot x) = 0 \qquad (vi)$$

(Factorization Preparation)

$$2\cdot\cos(2\cdot x)\cdot(\cos(x) + 1) = 0 \qquad (vii)$$

(Factorization)

$$\cos(2\cdot x) = 0 \ \lor \ \cos(x) + 1 = 0 \qquad (viii)$$

(Solving First Factor)

$$\cos(2\cdot x) = 0 \qquad (ix)$$

(Isolation)

$$x = 180\cdot n_1 + 45 \qquad (x)$$

(Solution)

(Solving Next Factor)

$$\cos(x) + 1 = 0 \qquad (xi)$$

(Isolation)

$$\cos(x) = -1 \qquad (xii)$$

(Isolation)

$$x = 180\cdot(2\cdot n_2 + 1) \qquad (xiii)$$

(Solution)

Figure 4-3: The Worked Example after Operator Identification

Now the Precondition Analysis phase begins. The method is essentially simple. Basically, the idea is to find the **major aim** of each step, i.e. to answer the question "Why was the step performed?" As indicated earlier, this analysis in expressed in terms of the satisfaction of preconditions of the subsequent steps. In the simplest case, the first method is applied to satisfy some preconditions of the second method so that that method can be applied.[19] In turn, the second method is applied in order to satisfy some preconditions of the third method etc. Finally, the last method is applied in order to produce a solution.

---

[19] Note that some of the preconditions of the second method may already be satisfied by the original state, so in general there is no need for the first method alone to satisfy *all* the preconditions of the second method.

The situation is not always that simple, three complications can occur.

1. Firstly, sometimes two or more methods need to applied before a certain method can apply. For example, suppose methods L, M and N are applied consecutively, L and M between them supplying the preconditions of N. In some cases, L and M can be applied in either order.[20] In such cases it is wrong to assume that the reason for applying L is to allow M to apply. LP can spot such cases, see section 4.4.2.5.

2. Another problem is caused by the limitations of the program's description space, it may sometimes be unable to describe the preconditions that are being satisfied. This is also discussed in section 4.4.2.5 below.

3. The final reason why the above description is too simple is that the worked example is presented linearly, whereas the underlying structure may be more complex, see section 4.3.1 above. The example has to be considered in sections. This situation is very common, and occurs in the example discussed below.

The first stage of the Precondition Analysis Phase is to find all applications of the key methods: those leading directly to solutions, and, in the case of LP, applications of Change of Unknown and Factorization. (See section 4.3.2.3).

As described in section 4.3.1, the example may be in several sections, each with an "end". This happens after a Factorization step for example. Each section is considered separately back to the "parent" step that formed it. In figure 4-3, (ix) to (x) forms one section, and lines (xi) to (xiii) form another. The parent step is (viii). These sections are very simple. Both sections consist of an application of Isolation.[21] As Isolation is a key method that leads directly to a solution, no complex analysis needs to be done on these sections. LP simply records that each section consists of an application of Isolation, the purpose of which is to solve the equation.

Now, we concentrate on the major part of the example, from line (v) to (viii). Line (viii) is obtained from line (vii) by Factorization.

As noted earlier, Factorization is always a desirable step, i.e. it is a key method. It splits the equation into two or more simpler equations. (In this case, the two factors are solved easily by Isolation.) Why could not Factorization be applied earlier, at line (vi) say? The answer concerns the preconditions of the Factorization method. For Factorization to be applicable, the equation must be of the form

$$e_1 \cdot e_2 \cdot \ldots \cdot e_n = 0.$$

The equation at line (vi) is not of this form. Obviously, the equation at line (vii) is of the correct form, as we know that Factorization was applied at that point. So, Factorization cannot be applied at line (vi), but it can be at line (vii). So, at least one effect of the transformation from line (vi) to (vii) is to make Factorization applicable. LP assumes that this is the **major aim** of the transformation, i.e. the Factorization Preparation step is applied to allow the application of Factorization. As the names imply, in this case LP's assumption is correct.

Note that *some* of the preconditions of Factorization are satisfied at line (vi), the right-hand side of

---

[20]If the methods had STRIPS type properties, this would be the case if the effects of L do not provide any of the preconditions of M, and similarly with L and M interchanged. However, this condition is not sufficient for the LP methods.

[21]In the second section, there appear to be two applications of Isolation. However, LP knows that Isolation can be done completely or in stages, and for the analysis it treats this section as containing one complete application.

the equation is 0. However, the left-hand side is not a product in the unknown. This is the missing precondition, i.e. line (vi) fails to satisfy the precondition **lhs-prod(X,Eqn)**.

Why could not Factorization Preparation (FP) be applied earlier, at (v)? Again, the answer is expressed in terms of preconditions. FP requires that the equation is of the form
$$e_1 + e_2 + \ldots + e_n = 0 \qquad \text{(xiv)}$$
where the e_i contain the unknown, and that all the e_i have a common factor containing the unknown.

The equation in line (v) satisfies the first part of the precondition, in that it consists of a sum of terms equated to 0. However, the terms do not have a common subterm, so FP cannot apply. LP assumes that the aim of the application of the new rule is to satisfy the missing precondition, i.e. it produces an equation of the correct form, with common subterms.

In general, the problem is to find the aim of the application of method M at line i, to produce line i + 1. As an example, we will use the step from (v) to (vi) in the figure 4-3,
$$\cos(x) + 2 \cdot \cos(2 \cdot x) + \cos(3 \cdot x) = 0$$
$$2 \cdot \cos(2 \cdot x) \cdot \cos(x) + 2 \cdot \cos(2 \cdot x) = 0$$
LP first finds the preconditions, P, of the method M_1 applied at line i + 1, to give line i + 2. In our case, M_1 is Factorization Preparation. The preconditions for FP are that the equation must be of the form (xiv) above, where the e_i must have a common subterm which contains the unknown.

LP represents this set P as follows:

1. The right-hand side of the equation is 0, written here as **rhs-zero(Eqn)**.

2. The left-hand side is a sum in the unknown, **lhs-sum(X,Eqn)**.

3. The equation contains top-level common subterms containing the unknown, X, written as **common-subterms(X,Eqn)**.

(Note that there is some redundancy in these conditions, the last condition in fact implies the second. However, the second condition is checked before the third, and it is a cheap test. If it fails, there is no need to check the third condition.)

LP then finds which of the members of P are satisfied at line i. Call these preconditions S. In our example, the first two elements of P are satisfied, but the last is not.

S is therefore
{rhs-zero(Eqn), lhs-sum(X,Eqn)}.
The set difference $P \setminus S$ is the set of preconditions not satisfied at line i, but satisfied at i + 1, call this set ME.[22] In our case, ME is
{common-subterms(X,Eqn)}.
If ME is non-empty, LP assumes that the aim of applying M is to satisfy ME so that M_1 can be applied. The set ME is called the **major effects** of the step, (the major aim of a step is to satisfy the major effects.)

If LP finds that all the preconditions of M_1 were satisfied, i.e. ME is empty, it looks for another

---

[22] ME must be satisfied at i + 1 or M_1 could not be applied.

explanation. This is described in section 4.4.2.5 below.

The above analysis is performed for each 'real' step in the worked example. (Real steps exclude those steps which are artifacts, such as partial applications of Isolation.)

The final analysis is shown in table 4-1. Note that Eqn in each line refers to the *current* equation, i.e. a different equation in each case. Similarly X refers to the current unknown, which happens to be x for every line. The first column corresponds to the line in figure 4-3. Note that not all lines appear, e.g. line (xii), those that do not are superfluous to the analysis. The last column of each of the last three entries is blank. In the case of the Isolation entries, each entry ends its section so no next method follows. Factorization is a key method, so no detailed analysis of the preconditions is required.

Line	Method	Major Effects (ME)	Satisfied Preconditions (S)
(vi)	Application of new rule	common-subterms(X,Eqn)	rhs-zero(Eqn) & lhs-sum(X,Eqn)
(vii)	Factor-ization Preparation	lhs-prod(X,Eqn)	rhs-zero(Eqn)
(viii)	Factor-ization	(Key Method)	-
(ix)	Isolation	(Solves Equation)	-
(x)	Isolation	(Solves Equation)	-

Table 4-1: Analysis of Worked Example

The analysis is used by LP to create new methods and schemas.

4.4.2.1 Creating New Methods

The next stage involves creating new operators, in the case of LP these are new methods. New methods are created if the program has been given new rules by the user. As LP is method-based, rather than rule-based, i.e. it learns *control* rather than factual knowledge, it must create new methods that apply the new rules. The new methods can be used in the same way as the original methods. By creating new operators, the programs increases its ability to solve problems.

A new rule was applied at line (v) in figure 4-3 above. The above analysis gives the preconditions of, and the aim of this step, expressed as the sets ME and S in table 4-1. LP creates a new method that can apply the rule. The new method allows LP to apply the rewrite rule whenever this is appropriate.

Finding the Preconditions

What are the preconditions of the new method? Basically, there is a choice. The preconditions can be obtained from the analysis above, or from the lines in the worked example, or from some combination of the two. LP adopts the first approach.

LP makes the preconditions of the method S. Hopefully, applying the operator will not undo the satisfaction of this set.[23] When LP eventually applies the new operator, if it succeeds in satisfying the major effect ME, and S is also satisfied, then *all* the preconditions of the following method M will be satisfied.

The postconditions of the method are the preconditions of the following method, M. When LP later applies the new operator, it will therefore test to see if the application of the operator does in fact preserve the satisfaction of S.

In this way, LP is able to create a method with preconditions that are probably necessary, although they are not sufficient. The preconditions are only probably necessary, because it may be that the rules used will automatically satisfy some of S anyway, even if these preconditions are not satisfied already.

In the example above, a new method is created that has as preconditions
{rhs-zero(Eqn), lhs-sum(X,Eqn), multiple-occ(X,Eqn)},
where multiple-occ(X,Eqn) is satisfied when there are multiple occurrences of X in Eqn.

The postconditions of the new method are the same as the preconditions, with the extra postcondition that the output equation must have common subterms, common-subterms(X,Eqn). These common subterms are needed for Factorization Preparation. LP records this fact along with the new method. We say that Factorization Preparation is the **indicated next method** of the new method. The indicated next method is used when solving new equations, see section 4.5.

Let us call the new method New.

4.4.2.2 Updating the Table of Connections

Whenever a new method is created, the Table of Connections is also updated. The table will probably already contain entries for at least some of the postconditions of the new method. The name of the new method must be added to each of these entries, (or new entries started) indicating that the correct application of this method *must* satisfy these conditions.

At present, LP does not update the other type of entry, indicating that a method may achieve the satisfaction of a condition. This would require the program to examine every application of a method, checking to see if it achieved effects that had not previously been recorded for that method. This facility could be added, but we have not felt the need to do so at present.

4.4.2.3 Choices in Applying a New Method

In general, it may be possible to apply the rule of the new method to an expression in several different ways.[24] For example, the rule of New above can be applied in three different ways to the

---

[23] LP checks for this, see next paragraph.

[24] Additional choices may arise because the method may have more than one rule, see the next section.

equation
$$\cos(x) + \cos(2 \cdot x) + \cos(3 \cdot x) = 0,$$

producing
$$2 \cdot \cos(3 \cdot x/2) \cdot \cos(x/2) + \cos(3 \cdot x) = 0,$$

$$\cos(x) + 2 \cdot \cos(5 \cdot x/2) \cdot \cos(x/2) = 0, \text{ and}$$

$$2 \cdot \cos(2 \cdot x) \cdot \cos(x) + \cos(2 \cdot x) = 0.$$

(These equations are produced by adding the first and second, second and third, and first and third cosine terms respectively.) However, the postconditions of New require that the resulting equation has common subterms. Only the third equation meets this requirement, and this is the only valid result of the application of New.

In general, LP applies the rule of a method in all possible ways until it finds one that meets the postconditions of the method.

4.4.2.4 Rule Lists

Each new method has a **rule list**, which is a list of rules that the method can apply. This corresponds to the partitioned rewrite sets of the PRESS methods. The rule list of a new method initially contains the new rule that was used in the example, so in the example a rule list is started for New. This rule list contains the rule (iv),
$$\cos(A) + \cos(B) \implies 2 \cdot \cos((A + B)/2) \cdot \cos((A - B)/2),$$
we will call this rule **cosrule**.

The rule lists are built up by the process described above in sections 4.4.1 and 4.4.1.2. Sometimes, during Operator Identification, LP discovers that a step satisfies the preconditions and postconditions of an existing new method New, but it finds that New cannot produce the required transformation. This implies that New is missing the relevant rule. In such cases, when the user supplies the rule, LP simply adds the rule to the set of rules used by New, by adding the rule to the rule list of New. This way, new methods can extend their sets of rules.[25]

Note that we could treat the new methods in exactly the same way as the existing methods by storing new rules in the set of a new method. However, the current implementation treats new rules slightly differently from existing rules.

During Operator Identification, if no existing *method* can explain a step, LP checks to see if the step uses a new rule (as part of the application of some as yet unknown method). This has been found to be convenient for our examples, as we often supply LP with several examples where one rule is used in different ways. The rule list approach means that we do not have to give the program the rule every time. This approach could be generalized for all the rules, so that LP would check to see if, for example, a Collection rule was being used in a new way. However, the current version does not do this.[26]

---

[25] The corresponding technique for extending the rule set of the original LP methods was described in section 4.4.1.2 above.

[26] Of course, the "converse" process is implemented. LP does check to see if new rules are being used in an old way, see section 4.4.1 above. Here, we are discussing old rules being used in new ways.

4.4.2.5 When the Set ME is empty

Sometimes, LP will find that the step it is trying to analyze satisfies no missing preconditions of the following method. In the terms used above, the set ME is empty. LP is able to recognize two cases where this happens.

Manipulating the Equation

One possibility is that the rule is used to manipulate the equation so that M_1 can be applied, although no new preconditions are satisfied. This kind of behaviour occurs because, in general, the preconditions are necessary but not sufficient conditions.

This can happen with the standard PRESS methods. Consider the following example:

$$\log_e(x + 1) + \log_e(x - 1) = 3 \qquad \text{(xv)}$$

$$\log_e((x + 1) \cdot (x - 1)) = 3 \qquad \text{(xvi)}$$

$$\log_e(x^2 - 1) = 3 \qquad \text{(xvii)}$$

Line (xvi) is obtained from (xv) by Attraction. Line (xvii) is obtained from line (xvi) by Collection. The precondition for Collection is merely that there are multiple occurrences of the unknown. However, this precondition is already satisfied at line (xv). The unknowns have been brought closer together by the Attraction step, but the preconditions of Collection do not contain facts about the unknown separation. The Collection rule used cannot apply unless the Attraction step is performed, but LP cannot express this in terms of satisfying preconditions. Here of course, LP can describe the difference between lines (xv) and (xvi), the unknowns in (xvi) are closer than the unknowns in (xv).

In general, the problem is harder, LP may actually lack the necessary terms in its description language. Consider the following worked example:[27]

$$x + \sec(2 \cdot x) - 1/\cos(2 \cdot x) = 0 \qquad \text{(xviii)}$$

$$x + 1/\cos(2 \cdot x) - 1/\cos(2 \cdot x) = 0 \qquad \text{(xix)}$$

$$x = 0 \qquad \text{(xx)}$$

The step from (xix) to (xx) is a Collection step. As before, all the preconditions of Collection are in fact satisfied at line (xviii), but Collection cannot be applied at line (xviii). LP cannot account for the step from (xviii) to (xx) in terms of preconditions, and moreover, it cannot describe the difference between the two lines in terms of preconditions. The required terms are outside LP's description language.[28] As LP knows that none of its existing methods account for this step, it needs to create a new method. In general, if a step from line i to line i + 1 does not satisfy any new preconditions of the method M_1 applied at line i + 1, no existing method can produce the step, and M_1 could not be applied at line i, then LP creates a new method M in the following way:

The user has already supplied the rule R that allows the step to be performed. If the preconditions of M_1 are P, the method M has P as *both* its preconditions and postconditions. M has R on its rule list, and M_1 is the indicated next method. The major effects of M are empty, and LP informs the

---

[27] While this example appears somewhat contrived, recall that the user can present any sort of example to LP.

[28] Work is being done on how new concepts can be added to the language automatically, see [79, 80, 56], and section 5.4.4.5. LP has no such ability at present.

user that this step does not appear to satisfy any new preconditions of the following method but it appears to be essential anyway. The fact that M has no major effects can be used when LP is solving new equations, see section 4.5.2.6.

Parallel Steps

The other case is where two (or more) steps have to be applied before a third, but the order of the first two is arbitrary. Consider the following example, which shows only the first four lines of a worked example.

$$\cos(4\cdot x) + \cos(6\cdot x) + \sin(4\cdot x) + \sin(6\cdot x) = 0 \tag{xxi}$$

$$2\cdot\cos(5\cdot x)\cdot\cos(x) + \sin(4\cdot x) + \sin(6\cdot x) = 0 \tag{xxii}$$

$$2\cdot\cos(5\cdot x)\cdot\cos(x) + 2\cdot\sin(5\cdot x)\cdot\cos(x) = 0 \tag{xxiii}$$

$$2\cdot\cos(x)\cdot(\sin(5\cdot x) + \cos(5\cdot x)) = 0 \tag{xxiv}$$

For this example, we assume that LP has just been given the rules that enable it to perform the steps from (xxi) to (xxii) to (xxiii), and it is about to create new methods for these steps.

Factorization Preparation is applied at line (xxiii) to produce line (xxiv). On first analysis, the aim of the step from line (xxii) to line (xxiii) is to satisfy the missing common subterms precondition of Factorization Preparation. LP creates a new method for this step that has the common subterms condition as a postcondition. When the first step is analyzed, this step satisfies no new preconditions, and this seems similar to the case discussed above. However, there is an important difference. In the examples discussed above, the Collection step could not be applied before the first step had been performed. Here, the rule used for the step from (xxii) to (xxiii) could be applied earlier. LP notes that the two rules can be applied in either order, and creates a method that applies both, telescoping the two steps.

4.4.2.6 The Division of the Worked Example

The next step involves dividing the worked examples into sections. This division is used to construct schemas. The first step is to give the worked example a **type**. The type of a schema depends on which key methods were used in the example.

At present, LP has three types.

If the worked example contains an application of Change of Unknown the type is **C.O.U.**, otherwise, if it contains a Factorization step its type is **Factorization**. If neither of these steps occur, the type is **General**. All worked examples eventually produce a solution, so those of type General contain a step that leads directly to a solution, Isolation or Polynomial Methods.

The worked example can now be divided into sections. In general, examples of different types are divided differently, and the program has to be told how to divide examples of each type. (This information can be considered as part of the knowledge needed by the program to utilize key methods, see section 4.3.2.3).

In the case of LP, if the example is of type C.O.U., it is divided into three sections. The first section contains the steps leading up to the Change of Unknown step, the second section contains the steps that solve the changed variable equation, and the final section contains the steps involved in the solution of the substitution equation.

If the example has type Factorization, and the Factorization step produces n factors, the example will contain n + 1 sections. The first section contains the steps leading up to the application of Factorization. The remaining sections each contain the steps involved in solving one factor.

If the example has type General, all the steps go into a single section.

Figure 4-4 shows the division of the worked example of figure 4-3 above. (Note that the application of the new rule is now described as an application of the new method New).

4.4.2.7 Schema Formation

Now most of the work has been done, and the schema can be created easily. The schema contains three main parts. These are

- the generating equation,
- the unknown that is being solved for,[29]
- and the **body** of the schema.

The body is a list consisting of all the operators used in the worked example. Each step is tagged with the following:

1. The operator used, (and additional identification information, e.g. LP requires the name of the rule used where applicable[30]).

2. The preconditions that it is used to satisfy, i.e. the major effects.

3. Any conditions that must also be maintained.

The schema is divided into the divisions that were found above.

Part of the schema produced by LP for the worked example of figure 4-2, is shown in figure 4-5. The column labelled SATISFIED PRECONDS shows which of the preconditions of the *following* method are already satisfied at that stage in the schema. The column labelled PURPOSE shows which of the preconditions of the following method need to be satisfied by the application of the method. The variables are not shared between the lines, thus Eqn stands for a different equation in each case. E1 is the result of applying the method to Eqn.

4.4.2.8 Storing the Schema

The schema is stored as a method. The preconditions of the method are the preconditions of the first method in the schema. The postconditions are that the equation is solved.[31]

These methods are called **schema methods**. LP will not create new schema methods unnecessarily. Before storing a schema, LP checks to see if it already has a similar one. A schema is similar to another one if they both contain the same steps, have the same type, and the preconditions

---

[29] In general domains, the part of the generating equation is played by the task solved by the example, the **generating task**, and the unknown is replaced by a (possibly empty) set of relevant parameters.

[30] Only user supplied rules have names, so the name is only applicable when these rules are used.

[31] In fact, the postconditions are not really used, see section 4.5.

```
┌─────────────────────────────────────────────────────────────┐
│                                                             │
│  cos(x) + 2·cos(2·x) + cos(3·x) = 0                         │
│                                                             │
│          (New method New)                                   │
│                                                             │
│  2·cos(2·x)·cos(x) + 2·cos(2·x) = 0                         │
│                                                             │
│          (Factorization Preparation)                        │
│                                          STEPS TOWARDS      │
│  2·cos(2·x)·(cos(x) + 1) = 0             FACTORIZATION      │
│                                                             │
│          (Factorization)                                    │
│                                                             │
│  cos(2·x)= 0 v cos(x) + 1 = 0                               │
│─────────────────────────────────────────────                │
│                                                             │
│          (Solving First Factor)                             │
│                                                             │
│  cos(2·x) = 0                                               │
│                                                             │
│          (Isolation)                                        │
│                                                             │
│  x = 180·n₁ + 45                         GENERAL PART       │
│                                                             │
│          (Solution)                                         │
│                                                             │
│─────────────────────────────────────────────                │
│                                                             │
│─────────────────────────────────────────────                │
│                                                             │
│          (Solving Next Factor)                              │
│                                                             │
│  cos(x) + 1 = 0                                             │
│                                                             │
│          (Isolation)                                        │
│                                                             │
│  cos(x) = -1                             GENERAL PART       │
│                                                             │
│          (Isolation)                                        │
│                                                             │
│  x = 180·(2·n₂ + 1)                                         │
│                                                             │
│          (Solution)                                         │
│                                                             │
└─────────────────────────────────────────────────────────────┘
```

Figure 4-4: The Division of the Worked Example

and purpose columns for each step are identical. If a similar schema already exists, LP does not store the new one. This is done to avoid unnecessary duplication of schemas. Storing duplicate schemas would slow down the process of solving equations. All schemas have to be examined, so the fewer there are the better. (The equation solving process is described in the next section.)

Schema methods have a slightly special status. When LP is solving a new equation, it tries to find a

NAME	SATISFIED PRECONDS	PURPOSE
Method New (cosrule)	{rhs-zero(Eqn), lhs-sum(X,Eqn), multiple-occ(X,Eqn)}	common-subterms(X,E1)
Factorization Preparation	{rhs-zero(Eqn), lhs-sum(X,Eqn), multiple-occ(X,Eqn), common-subterms(X,Eqn)}	lhs-product(X,E1)
Factor- ization	{rhs-zero(Eqn), lhs-prod(X,Eqn), multiple-occ(X,Eqn)}	(Major Step)
Isolation	{single-occ(X,Eqn)}	(Solution)
Isolation	{single-occ(X,Eqn)}	(Solution)
Generating Equation: cos(x) + 2·cos(2·x) + cos(3·x) = 0 Unknown: x Schema Type: Factorization		

Figure 4-5: Part of the Schema

suitable schema method before it tries any other methods, see section 4.5.

Also, when examining worked examples, LP does not consider the schema methods during Operator Identification. If schema methods were used in the worked example, the example would not really be comprehensible, since these methods can produce huge "jumps" if no intermediate steps are presented. For instance, consider the schema method produced from the schema of figure 4-5. Using this method can produce the two line worked example

$$\cos(x) + 2\cdot\cos(2\cdot x) + \cos(3\cdot x) = 0$$

$x = 180\cdot n_1 + 45 \quad \lor \quad x = 180\cdot(2\cdot n_2 + 1),$

which is not easily understood![32]

4.4.2.9 The Strategy of the Schema

The schema "summarizes" the worked example. However, it is an incomplete summary. Section 4.2.1.2 described what a student could learn from the worked example. As remarked there, a deep analysis shows that the method demonstrated in our standard worked example works on equations of the form

---

[32] If the user presents an example using one, LP fails to understand the step as the application of the method, and processes it in the normal way.

$$a \cdot \cos(p) + b \cdot \cos(q) + a \cdot \cos(r) = 0, \qquad \text{(xxv)}$$
where $q = (p + r)/2$.

However, the schema does not really reflect all of this. LP may attempt to use the schema on any equation which satisfies the preconditions of the schema: that the left-hand side is a sum, the right-hand side is 0, and there are multiple occurrences of the unknown. There are obviously many equations of this form for which the schema is unsuitable.

If more than one schema method has its preconditions satisfied, LP uses an extra test to choose between them (see section 4.5.1 below). The type of equation that matches best with the schema method above is of the form

$a \cdot \text{trig}_1(p) + b \cdot \text{trig}_2(q) + c \cdot \text{trig}_3(r) = 0$,

where the trig_1 are sines or cosines, and p, q and r are arbitrary angles. This equation is more general than equation (xxv) above, and does not satisfy all the conditions satisfied by equation (xxv).

If the extra conditions are not met, the schema cannot be used exactly, i.e. there will come a point when the next method listed in the schema cannot be applied. In this case, LP may be able to "patch" the schema (see section 4.5.2). In general, the task of finding exactly the right conditions is very difficult, and either requires a great deal of knowledge or some form of concept-learning, see [53] and section 4.7.1. As the schema is executed in a flexible way, LP does not need to tackle this task fully. Instead, LP does a partial (precondition) analysis, and tries to patch the plan if necessary. This seems to be a reasonable compromise in the equation solving domain, and probably applies to many other domains.

4.4.3 Plan Recognition

As described earlier, Operator Identification is somewhat similar to proof checking. Similarly, the Precondition Analysis Phase resembles **Plan Recognition**.

Plan recognition is the task of inferring the plan of an entity, the **actor**, from a sequence of the actor's actions. The entity that infers the plan is usually called the **observer**.

Much of the plan recognition work in A.I. focuses on the problem of understanding natural language discourse, e.g. [2, 86]. Each speaker has to recognize the purpose behind utterances of the other speaker(s), and adjust the dialogue accordingly.

Plan recognition has some obvious similarities with Precondition Analysis. LP has to recognize the purpose of each step in a worked example, provided by another entity. However, there are some important differences:

- The two processes both infer plans, but the plans are used for different purposes. In Precondition Analysis the task is to infer the plan *and use it for future problem solving*. The plan may need to be generalized in some way so that it can be used for problem solving. In Plan Recognition, once the plan has been identified, the task is complete.

- In Plan Recognition, the task of the observer is to infer the goal (and plan) of the actor. In the case of Precondition Analysis, the goal is already known (i.e. to solve the equation).

- Another important difference is that in Plan Recognition the observer (usually) knows the preconditions and effects of all of the actor's actions, the problem is to see how the actions fit together. In contrast, Precondition Analysis is able to learn *new* operators and assimilate them into its plans.

It seems that Precondition Analysis and Plan Recognition are used for tasks which differ sufficiently to prevent techniques used in one being of much use to the other. Certainly, we have found no new techniques in the Plan Recognition literature that were of help to our work here, and the converse may well be true!

4.5 Solving New Tasks

This section describes in more detail how the schema is used to solve new tasks. If the program has previously analyzed some examples, it tries to find a schema method that will help it solve the problem. The schema gives direction to the problem solving attempt, and its use shows one way in which the program has learned from the worked examples. However, the program may have created new methods from other examples, and these can be important too.

Sometimes, the program will not be able to find a suitable schema, and it may then try to solve the problem using standard techniques. In the case of LP, if no schema can be found LP uses techniques similar to those of PRESS, see section 4.5.3.

The next section describes how the program chooses a schema, and the following sections describe how the schema is used to guide the problem solving process.

4.5.1 Choosing the Appropriate Schema

The user gives the program a task to solve, we call this task the given task. (In the case of LP, the **giveind equation**). The program first tries to find a schema method that might help it solve the given problem. To do this, the program finds a schema method whose preconditions are satisfied by the given task. This ensures that at least the first operator in the schema *might* be applicable (recall that the preconditions are necessary but not sufficient). If the preconditions are not satisfied, then as the first operator in the schema certainly cannot be applied, there is no reason to believe that the schema method can be used to help solve the equation.

The process of checking the preconditions of the schema operator against the given task that is to be solved is a crude analysis, but it is a reasonable compromise. One alternative approach would be to have the program perform an extensive analysis of the steps in the schema, and decide if they were appropriate to the new situation. Such an analysis would obviously expend a large amount of resources, and the effort may well be wasted. LP adopts the "quick and dirty" approach, i.e. it checks that the given equation satisfies the preconditions of the schema method and then it attempts to use the schema. It does not spend time making sure that the schema is suitable.[33]

Sometimes, the given task will satisfy the preconditions of more than one schema method. In such cases, the program can use some sort of procedure to choose which to try first. LP selects those schemas which satisfy an extra condition, those with a generating equation whose functional type is the same as that of the given equation.[34] Why is this a good test?

---

[33] Of course, similar remarks apply to the way ordinary methods are chosen by PRESS (and LP when it is not guided). PRESS just checks to see if the preconditions are satisfied and then tries to apply the method.

[34] If more than one schema passes this test, what should LP do? One possibility is for LP to select one arbitrarily. Alternatively, it can apply more tests, such as choosing one with whose generating equation has the same number of unknowns as the given equation. The current version of LP allows the user to decide which of these courses should be followed. In the latter case, LP can apply a series of tests. These are rather ad-hoc, so they are not described here.

If the generating task and given task are identical, the schema method is certainly appropriate! If the two tasks differ greatly, the schema is less likely to be useful. Despite the fact that the relation is far from exact, we feel that the schema is more likely to be appropriate the more closely the two tasks resemble each other. Note that the generating task must have satisfied the preconditions of the first operator. By insisting that the given task also satisfies these preconditions, we ensure that the two tasks are at least slightly alike. In some cases, however, these preconditions can be very weak, and the two tasks may differ greatly. The functional type test is an extra way of ensuring the similarity of the tasks.

Suppose that the first method in a schema is one with weak preconditions, e.g. Collection, (precondition multiple-occ(X,Eqn)), but the following methods require the equation to be trigonometric. The extra test prevents LP trying to use the schema on a logarithmic (say) equation with two occurrences of the unknown. Again, this approach is quick and dirty. It is "dirty" because a schema that fails this test may be perfectly applicable, if it uses methods that do not in fact rely on any functional type. Also, many schemas that pass this test may be inapplicable later on. However, these failings are not very important because this test is only used for schema ordering, all schemas whose preconditions are satisfied can be tried if necessary.

The test is "quick" computationally, and experimentally seems to produce good results. It could be extended to check for other features, but this has not proved necessary.

4.5.1.1 Choosing a Subschema

If the schema is of type Factorization, and eventually Factorization is successfully applied, the program will have several factors to solve. Corresponding to these factors are General type sections of the schema. LP has to decide which of these schema sections should be used for each factor. The process is the same as that described above for the initial choice of schema, the current factor must satisfy the preconditions of the first method used in the subschema. LP does not insist on a one-to-one correspondence between factors and sections, that would be unrealistic considering the laxity of the schema selection conditions, the generating and given equations can be fairly different. This means that sections that have been used once are not discarded, the same section can be used for all the factors if it provides the best choice for each.

4.5.2 Following The Schema

Once the schema has been selected, LP uses it to guide the attempt to solve the given equation. The schema lists the methods used to solve the generating equation, together with other information about preconditions and purposes (see section 4.4.2.7 above).

In essence, LP tries to apply the methods listed in the schema to the given equation. LP tries to use the schema from the top, i.e. it tries to apply methods in the same order as in the schema. LP appears to works down the schema, at any point it is working on a particular step of the schema. The method in this step is called the **current indicated method**, and we will say that LP is trying to apply the current indicated method.

If it succeeds in applying the current indicated method, LP moves on to the next step, and then tries to apply the new schema method, and so on. By trying to apply the steps it learned from the worked example in exactly the same sequence, LP is using the results of a form of learning called **learning by rote**. This is a reasonable first strategy: the steps solved the generating equation, and the given equation and generating equation both satisfy the preconditions of the first method in the schema. If the given equation is very similar to the generating equation, the attempt may succeed. More usually, there will come a point when a method listed in the schema cannot be applied to the transformed given equation.

If LP had learnt only the list of steps, it would now be stuck. This illustrates the weakness of learning by rote. It is a very weak learning technique because there is no understanding of the reasons for various decisions, so it is very difficult to recover from unexpected failure.

However, the schema contains other information as well. The most important parts are the division of the schema into sections, and the major aim of each step. This enables LP to modify the linear execution of the schema.

Basically, LP uses the flow-chart shown in figure 4-6, which shows how LP follows the chosen schema. The flow chart contains six "tests", some of which, such as "Apply C.I.M." are also actions in that they affect the current state.

Note that the first two tests allow LP to omit large numbers of steps. These tests look for states that are unexpectedly "good". Placing these tests first allows LP to take advantage of such fortuitous circumstances. The planner of STRIPS, PLANEX, [36], also does this, but there are differences. PLANEX always tries to apply the last step in a plan, then the last two etc. This allows STRIPS to execute the smallest part of the plan that is applicable. Such an approach is not useful for LP, as the operators are not well-behaved. See section 5.3 for further explanation.

The third test is the obvious one, try to apply the same operator that was used in the worked example.

The fourth test allows LP to replace one operator in the plan with another that achieves the same effect. If such an operator can be found, things may still go wrong, as the original operator may have had an undetected effect, but this substitution action is a reasonable heuristic.

The fifth test is less directed. LP cannot apply the operator it wants to, and it cannot substitute for the operator. LP tries to find an operator that will not undo any already satisfied preconditions, but which might hopefully manipulate the equation in some unspecified way, so that the desired operator may then be applicable. LP would not need to perform such actions if the operators were better behaved. Note that this procedure can be applied over and over again, allowing LP to *add* an indefinite number of steps to the plan.

The sixth test allows LP to omit an operator if it has no major effects. This is risky, as the operator did something in the original worked example, LP just could not describe what! Nevertheless, as a last resort, this method is reasonable.

Some details have been omitted, these are discussed in the sections below.

Section 4.5.4 contains examples of the various types of modifications.

4.5.2.1 Is the Equation Solved?

Before trying to apply a step, LP tests to see if the equation is completely solved. By completely solved, we mean that the equation is of the form $x = a$, (where a is free of x) or is a disjunct of such terms, and that no more equations (sub-problems) remain unsolved. So, an equation solved using Change of Unknown is not completely solved if only the changed variable equation is solved, the substitution equation must be solved as well.

If the equation is completely solved, LP can exit with success, the problem has been done. Otherwise, LP continues.

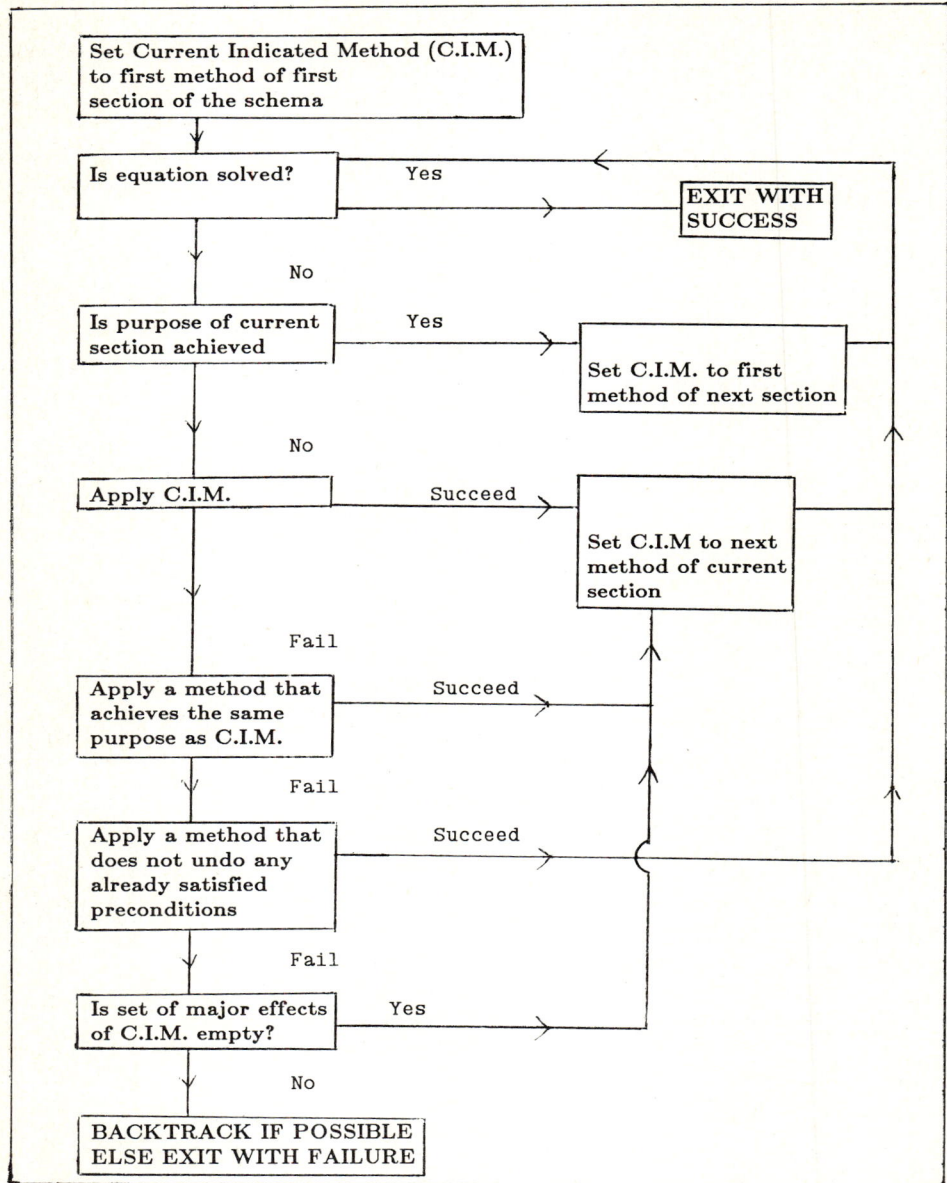

Figure 4-6: Flow Chart for Following Schema

4.5.2.2 Achieving the Purpose of a Section

As described above, section 4.4.2.6, the worked example and schema can be divided into sections, each section having a purpose.

The current indicated method must fall into a section of the schema. If the schema is a Factorization or C.O.U. one, the section is a proper subsection of the schema, otherwise the section is the whole schema. The section that the current indicated method is in is called the **current schema section**. Before trying to apply the current indicated method, (and after checking to see if the equation has been solved), LP checks to see if the purpose of the current section has already been achieved. If it has, then LP jumps all the remaining steps listed in that section, and moves on to the beginning of the next section.[35] This allows LP to omit an unlimited number of steps.

The purpose of a section is one of the following:

1. To allow Factorization to apply.

2. To allow Change of Unknown to apply.

3. To solve the equation not using one of the above methods.

LP can easily check to see if the purpose of a section has been attained. In the first case above, Factorization can be applied if the equation is a product equated to 0, i.e. of the form
$$e_1 \cdot e_2 \cdot \ldots \cdot e_n = 0.$$
Change of Unknown can be applied if all occurrences of the unknown occur within common subterms (and that this subterm occurs at least twice in the equation, and is different from the unknown, see section 2.3.1.7).

Finally, the purpose of the third type of section has been attained if the equation is solved, but this case is covered by the first test above.

4.5.2.3 Applying the Current Indicated Method

If the equation is not solved, and the purpose of the section has not been achieved, LP tries to apply the current indicated method. If this succeeds, the schema is working well, and LP continues its execution of the schema by making the next method the current indicated method and recursing. (i.e. LP now tests to see if the new state of the equation is solved. If it is not, it checks to see if the purpose of the current section is achieved etc.)

Suppose that the current indicated method is the last method in a section, does LP advance to the next section if it succeeds in applying the current indicated method? Yes, it does, but not because of this test. The successful application of the last method in a section *must* achieve the purpose, by construction, so this case is covered by a test higher up.

4.5.2.4 Substituting Operators

Achieving the purpose of a section is one way that the program can modify the linear execution of the schema. Another way makes use of the information the schema contains on the major aim of each step.

Suppose that the current indicated method is CIM, and LP finds that CIM cannot be applied to the current state of the given equation. Assume also that the purpose of the current section has not been achieved, and that the equation is not already completely solved (if either of these conditions held, LP would not have bothered to try to apply CIM.) In this case, LP examines the major effects listed for the CIM step. Suppose that the major effects of this step are ME.

---

[35]If the schema is a Factorization one, LP must choose which section of the schema should be used for each factor. See section 4.5.1.1.

Sometimes, as was seen in section 4.4.2.5 above, ME may be empty. See section 4.5.2.6 for this case. Here, we assume that ME is non-empty, i.e. the current indicated method has a major aim.

LP then looks for another method that might satisfy ME. This information is found by examining the Table of Conditions, (described in section 4.3.4). For each condition C in ME, LP first examines the table to find which methods *must* achieve the satisfaction of C. If one method must satisfy all the conditions in ME, LP considers this method first. Otherwise, LP looks for methods that potentially can satisfy all of ME.

In general, such a method will have a mixture of *must* satisfy and *may* satisfy entries, i.e. the conditions in ME are covered to differing extents. If several methods could potentially satisfy all of ME, LP prefers the one with most *must* entries (breaking ties arbitrarily). Actually, this feature is very rarely used, as much of the time ME is a singleton!

Other, more sophisticated, techniques can be used. LP could check that the method it chooses will not possibly destroy the satisfaction of the preconditions of the following method in the schema (using both the Connection and Condition tables), but the current implementation does not do this. After all, there may be no need to apply the following method, it may have no major effect. Also, we may not succeed in applying the method we are currently considering. There is a trade off between using a large amount of inference or just trying a method to see if it works. As elsewhere, because the operators of LP are not well-behaved, we choose to limit the amount of inference used.

If it finds a method that may be suitable, M say, LP tries to apply it instead of CIM. If M can be successfully applied, then its postconditions are satisfied, so, (by construction!) ME is satisfied, i.e. the major aim of the schema step has been attained. In this case, LP goes on to the next step in the schema, having substituted M for CIM. This approach is not trouble free. Recall that LP's analysis of the original example was limited, and it makes assumptions, due to its limited vocabulary, that may not be valid. Method CIM may have some effects that LP does not detect, so the substitution of M may not *really* replace CIM. However, this approach is a reasonable approximation.

If method M cannot be applied successfully, LP tries to find another possible method. If eventually, no more methods are potentially suitable, LP is unable to directly substitute for the step, and another approach, described in the next section, must be used.

Note that all members of ME must be satisfied[36], as M is substituted for CIM. If no such method can be found, one possibility is to look for methods that satisfy parts of ME, and apply them until all members of ME are satisfied. However, this causes all the standard planning problems of interactions, an attempt to satisfy one part may undo the satisfaction of other parts. Rather than attempting such an approach, LP insists that the only methods attempted are those that satisfy *all* of ME.

Restrictions

LP will not substitute for Factorization or Change of Unknown. The use of these methods are vital in the schema, and the structure of the schema reflects this.

Similarly, LP will not consider either of these methods as potential candidates for M (the method that is substituted for the current indicated method.) Using these methods radically changes the problem, as they introduce extra subgoals (e.g. solving the factors, or solving the substitution equation), whereas the CIM does not do this. LP would not be able to know which parts of the schema method should be devoted to each subgoal, and the whole plan of the schema would probably

---

[36] However, ME is usually a singleton anyway.

be inapplicable.

Similar restrictions apply to the remaining modifications as well.

4.5.2.5 Adding Steps

If the current indicated method cannot be applied to the current state of the equation, and the process described in the previous section fails, LP tries applying additional methods in the hope that these will transform the equation sufficiently so that the current indicated method can eventually be applied.

There is no good reason to assume that this approach will be successful, and if it was not carefully controlled LP would risk expending a great deal of effort uselessly. LP considers all of its methods to see which might prove useful, but it in fact only attempts to apply a small fraction.

Firstly, LP considers only non-schema methods. There is no point in trying to use a schema, it is already using one!

Also, the normal process of checking that the preconditions of a method are satisfied by the current state of the given equation cuts down the number of methods that LP need try. Also, LP does not try the current indicated method, as the attempt to use that has already failed!

A more important consideration concerns the preconditions of the current indicated method. There are two types of pruning operations.

Firstly, LP only considers those methods whose postconditions are compatible with the preconditions of the current indicated method. This ensures that if such a method is used, it is at least possible that the current indicated method would be immediately applicable.[37]

The other pruning operation occurs because some of the preconditions of the current indicated method will already be satisfied by the current state of the equation. LP will not apply a method that will undo such preconditions.

Both of these pruning operations use the Tables of Conditions and Connections. The Table of Conditions indicates which conditions are compatible (resp. exclusive) with the required conditions. The Table of Connections indicates which methods will cause the compatible (resp. exclusive) conditions to be satisfied, and these methods are considered (resp. excluded).

This generally leaves very few possible methods. If any methods do remain, LP tries each in turn (using the priority ordering to decide which to try next) trying to find one that can be applied. If such a method is found, LP behaves as if it had just finished executing the *previous* method, i.e. it performs all the steps described from section 4.5.2.2 above on the new state of the equation. So, it first checks to see if the equation is completely solved, if it is not it tests to see if the purpose of the current section has been achieved. If it has not, it tries to apply the current indicated step, which may now be possible. If this attempt fails, it now tries to substitute for the step (as described in section 4.5.2.4). Although this failed last time, it may succeed now because the current state of the equation is different. If this still proves impossible, LP can try to transform the equation again, using the technique described in this section.

Note that, in theory, this process allows an indefinite number of operator applications between steps

---

[37]This is a weak type of means-end analysis.

in the schema. Although this does not appear to cause difficulties with LP, there may be problems in some domains. One ad-hoc patch is to put a limit on the number of operator applications that can be applied between steps. An early version of LP adopted this approach.

This process of adding extra operators may fail, eventually, LP may find that there are no remaining methods that can be applied. The next three sections describe what happens in this case.

4.5.2.6 Omitting Steps

The step applying CIM in the schema may be omitted if the major effects set ME is empty. (See section 4.4.2.5 for examples where ME is empty.) If all the relevant attempts above[38] have failed, and ME is empty, LP skips that step and proceeds with the next step in that schema. The idea is obvious: no good reason has been found for the step, so the solution may be able to proceed without it.

However, it is not necessarily safe to omit the step. The reason is the same as mentioned above in section 4.5.2.4. LP's analysis is not perfect, and it makes assumptions, due to its limited vocabulary, that may not be valid. Method CIM may have some effects that LP does not detect, and omitting the step will not produce these effects. This is why steps are omitted only if the other approaches have failed.

Omitting a step is thus a last-ditch attempt to use the schema. LP cannot judge the success or failure of this move, as there are no clear expectations as to what should happen.

4.5.2.7 Starting Again

The whole attempt to use the schema finally fails when all the above modifications have been exhausted, and the CIM has a non-empty ME. In this case CIM cannot be omitted. LP tries to find another applicable schema method. If it finds such a method, it attempts to use that schema, starting from the *original* given equation. (The program starts from the original equation because the previous attempt may have gone down a "dead end". The situation is similar to the no schema case, described in section 4.5.3.1.) If there are no more untried applicable schema methods, the program attempts to solve the equation without the schema, this is described in section 4.5.3.

Another possible approach, not used by LP, would be to allow an operator to be omitted even if its ME was non-empty. (Presumably, this would only be used after operators with empty MEs had been omitted.) This approach is really last-ditch, the whole analysis is invalidated if operators with major effects are left out. However, if there is no other possibility, it is a reasonable thing to do, on the assumption that perhaps bits of the schema might still be useful. In the case of LP, there is another possibility, described in the next section.

4.5.3 Solving Problems without the Schema

This section describes how LP solves equations without a schema. Basically, in this situation LP behaves like PRESS, except that, at each stage, LP tries to find a schema that it can use, see section 4.5.3.2. The waterfall of LP is very similar to that of PRESS, (see section 2.3.3), the difference being that program-created methods have to be fitted in.

Schema methods are at the top of the waterfall, the reason for this is described in section 4.5.3.2 below. Other program created methods are tried after the key methods, Isolation, Polynomial

---

[38] i.e. achieving the purpose of a section and applying the method directly.

Methods, Change of Unknown and Factorization, but before the remaining methods.[39]

Why should LP attempt to solve an equation without a schema? This situation can arise in two ways. The first possibility is that LP may not have a schema that it can use. This can happen because it has no schema methods or because the given equation does not satisfy the preconditions of any schema method.

The other case occurs when LP has been following a schema but the current step cannot be applied, and all the possible actions above have been exhausted, and no more applicable schema methods remain untried. The step that fails may not be the first step in the schema. In this case, the given equation may have been extensively transformed by the earlier steps in the schema. Nevertheless, LP ignores all the transformations up to this point, and attempts to solve the given equation from scratch without the schema.

4.5.3.1 Why Return to the Given Equation?

It may seem strange to lose all the work done up to this point, why not ignore the schema but carry on from the current state of the equation? The answer is that the previous transformations may be unhelpful, leading to a dead end. If the schema analysis breaks down at a certain point, there is no reason to assume that the previous steps should have been applied. In particular, without the schema, LP would use the priority ordering of the methods implemented by the waterfall (see section 2.3.3). The default ordering combines some obvious relative orderings, (key methods first) with some that were determined experimentally, and the default ordering performs better than any other ordering we tried. The use of the schema totally overrides the ordering, so that obvious steps may not be attempted. This may cause the current equation to be considerably more complex that the given equation. Once the schema has failed, it is better to start again, so that the waterfall has a chance to operate fully on the original given equation.[40]

4.5.3.2 Finding a Schema

Although LP can operate without a schema, it is preferable for it to be guided. The schema methods are placed at the top of the waterfall. If LP is not being guided by a schema, after every transformation of the equation, it will check to see if it can use a schema method now. If a suitable schema method is found, LP tries to use the schema as described above.

Suppose that LP is given the equation G, and is unable to solve it using a schema method, either because no schema methods exist, or all they fail when tried. It then tries to transform G using the methods in the waterfall. Suppose it transforms to G to G'. The equation G' is passed over the top of the waterfall, and is processed by the schema methods first. LP can use a schema method if its preconditions are satisfied by G'.[41] At this stage, it is as if LP had been given the equation G' originally. If the schema fails (after all modifications have been tried), LP returns to G' and tries again. If no schema method works, G' continues down the waterfall, and is processed by the other methods. (Note that backtracking may eventually cause LP to return to G, and cause it to try and

---

[39] In order to allow experiments, LP allows the user to change the relative ordering of the lower-level methods. Also, various methods can be disabled, if this is desired. The default ordering is the same as in PRESS.

[40] LP could be extended so that it firstly tried to solve the equation from the current state, only returning to the original equation if this attempt failed.

[41] In fact, LP will not use a schema method that has already failed on G. This is to avoid the possible duplication of work: when trying to use the schema on G, it is quite likely that one of the modifications produced G', so there is no point in trying again. Note that this is a heuristic, the schema method may in fact work on G'. However, even if this is the case, LP may be able to find the solution anyway, and this restriction is a quick and simple way of avoiding possible duplication.

find another way of transforming G.)

4.5.3.3 Differences due to the Lack of a Schema

What difference does the lack of a schema make to the behaviour of LP? The first obvious difference is that LP has no explicit plan. However, various parts of schema execution are implemented by the waterfall and by the new methods.

- In the default ordering, the key methods are at the top of the waterfall (after the new methods), i.e. LP attempts to use them first. If these methods can apply, LP will use them. This is somewhat similar to achieving the purpose of a section of the schema, the first thing that LP checks for when executing a schema.

- Many new methods contain a suggestion as to which method should be tried next, the indicated next method. The indicated next method is the one whose preconditions are satisfied by the major effects of the new method (see section 4.4.2.1). This is usually the method that was applied next in the worked example that generated the new method. When no schema is used, LP always tries the indicated next method first, so the waterfall is overridden to some extent.

- As described above, LP tries to find a schema method before trying anything else in the waterfall.

Another difference caused by the lack of a schema is in the backtracking behaviour. The backtracking of PRESS was described in section 2.3.3.1 above, and LP behaves in a similar way without schemas. LP does not backtrack in this way if a schema is being used. The schema guides the program so there is less need for backtracking in this phase, if things go wrong a second attempt will be made without the schema.

4.5.3.4 Reporting Success or Failure

Whenever LP solves an equation, whether it uses a schema or not, it reports the answer. Sometimes, however, more information is given.

If LP succeeds in solving an equation with a schema method, it informs the user that there were no problems with the schema. If LP tries to use a schema method, and all the schemas fail to help LP solve the equation, but LP then solves the equation *without* using schemas, LP issues a warning to the user. The idea behind the warning is that it would have been better if the schema methods had not been used. A full self-improving algebra system, (see section 1.5.2.1), would make use of this warning by either adding extra preconditions to the schema methods using concept-learning techniques, or by modifying the schemas in some way. LP has no such ability at present, and LP merely notes that this event has occurred. It stores the solution trace and the schema methods responsible. These could be passed on to another component in the full system or LP could be extended to deal with this, see section 4.7.1.

If the equation cannot be solved by LP, either with or without a schema method, LP informs the user of failure. No special action is needed in this case as there is no reason to assume that the schema methods used, if any, are faulty. There are some problems that LP just cannot solve!

4.5.4 Examples of Schema Modification

This section contains examples of some of the schema-modifying steps used by LP. In the sections above, the modifications were discussed in the order that LP considers them. Here, the examples are presented in decreasing order of frequency of occurrence in our experience. Schemas are modified more often by adding steps than omitting steps, and both modifications are more common than the substitution of steps. The first modification described above, achieving the purpose of a section, occurs rarely on our examples.

4.5.4.1 Adding Steps

As an example, suppose that LP is given the equation
$$2\cdot\cos(x) + 3\cdot\cos(3\cdot x) + 2\cdot\cos(5\cdot x) = 0 \qquad \text{(xxvi)}$$

to solve, and that it chooses to use the schema shown above in figure 4-5. This schema was generated by the equation
$$\cos(x) + 2\cdot\cos(2\cdot x) + \cos(3\cdot x) = 0,$$

and is a Factorization type schema.

The first step is to see if the equation is completely solved. It is not, so the next step is to see if the purpose of the current schema section has been attained. The purpose of the current section is to allow Factorization to apply. Equation (xxvi) is not a product, so Factorization cannot be applied, so the purpose of the current section has not been attained yet. So, LP attempts to apply the current indicated method.

The first method in the schema is New, which applies the rule
$$\cos(A) + \cos(B) \Longrightarrow 2\cdot\cos((A + B)/2)\cdot\cos((A - B)/2),$$
in order to provide common subterms for Factorization Preparation.

The rule cannot be applied directly to equation (xxvi), as the cosine terms have multiplicative factors dominating them. The attempt to apply New to this equation therefore fails.

LP now tries to find a method that would achieve the major effects of New, i.e. it tries to find a method that produces common subterms for Factorization Preparation. Suppose that it cannot find such a method.[42] LP now tries to use other methods, in the hope that it will then be able to apply New. These methods must not alter any preconditions of New that are already satisfied by the equation (xxvi).

There are three such preconditions: the right-hand side is 0, the left-hand side is a sum, and contains multiple occurrences of the unknown.

Many methods can be ruled out quickly. The Table of Connections shows that Isolation and Polynomial Methods solve the equation, and the Table of Conditions show that this condition excludes the multiple occurrences condition. (Solutions count as only one occurrence of the unknown.)

Factorization and Change of Unknown are not considered, as they produce extra subgoals. (Also, Factorization need not be tried because LP has already tested to see if the purpose of the current section has been attained.)

---

[42] This will be the case if LP has seen only the one worked example before being given the new equation. Other worked examples might cause LP to create other methods with this major effect.

The current equation does not satisfy the preconditions of Factorization Preparation, Homogenization, Logarithmic Method, and Nasty Function Methods, so these methods are ruled out by the precondition check.

This leaves two methods that must be considered fully, Attraction and Collection.[43] The order in which the two methods are tried is determined by the priority ordering. With the default ordering, Collection is tried first. This fails (no Collection rule can transform the equation) so Attraction is tried. This transforms equation (xxvi) to

$$2 \cdot (\cos(x) + \cos(5 \cdot x)) + 3 \cdot \cos(3 \cdot x) = 0. \tag{xxvii}$$

The equation has been transformed. LP first checks to see if the equation is solved. It is not, so LP tests to see if the purpose of the current schema section has been attained. In this case, as Factorization cannot be applied to equation (xxvii), the purpose still has not been achieved. LP now tries to apply the current indicated method, New. This attempt succeeds on the new equation, (xxvii), and from then on the schema can be applied exactly. Appendix V shows the output produced by LP solving equation (xxvi).

4.5.4.2 Omitting Steps

Consider what happens if the given equation and generating equation are interchanged in the example above. In this case, LP is shown a worked example for the equation (xxvi). This produces the same schema as before, except that there is an additional step at the beginning, an application of Attraction. The ME of this first step is empty, the Attraction step appears to satisfy no new preconditions for the application of New, but New cannot be applied unless this step is performed first (this is similar to the first example in section 4.4.2.5).

Suppose now that this schema is used to solve the old generating equation

$$\cos(x) + 2 \cdot \cos(2 \cdot x) + \cos(3 \cdot x) = 0.$$

The equation is not already solved, so as before, LP checks to see if the purpose of the schema has been attained. It has not, so LP tries to apply the current indicated method, Attraction. This attempt fails. As the ME of this step is empty, LP assumes that the step can be omitted. LP therefore attempts to apply the next method in the schema, New. This succeeds, and all the following steps go through.

4.5.4.3 Substituting Steps

Suppose that LP is given the equation

$$\cos(x) + 2 \cdot \cos(x) + \cos(x) = 0 \tag{xxviii}$$

to solve, and that it uses the schema used in the first example above, section 4.5.4.1(i.e. the schema shown above in figure 4-5. This schema was generated by the equation

$$\cos(x) + 2 \cdot \cos(2 \cdot x) + \cos(3 \cdot x) = 0,$$

and is a Factorization type schema.)

As before, the first step is to see if the equation is completely solved. It is not, and the next test is to check if the purpose of the current schema section has been attained. Equation (xxviii) is not a product, so Factorization cannot be applied, so the purpose of the current section has not been attained yet. So, LP attempts to apply the current indicated method.

---

[43]Collection cannot be ruled out just because it reduces the number of occurrences of the unknown. It does not necessarily reduce it to a single occurrence, it can reduce from three occurrences to two for example.

The first method in the schema is New, which applies the rule
$$\cos(A) + \cos(B) \Longrightarrow 2\cdot\cos((A + B)/2)\cdot\cos((A - B)/2),$$
in order to provide common subterms for Factorization Preparation.

The rule cannot be applied to equation (xxviii) in such a way as to produce common subterms, so the attempt to apply the current indicated method fails. LP tries to find a method that will achieve the effect of New, i.e. a method that will produce common subterms for Factorization Preparation. It looks in the Table of Connections and discovers that the only method which *must* produce common subterms is New. Obviously, there is no point in trying this method again, so it looks for methods that *may* produce common subterms. Collection and Attraction are such methods.

In this case, Collection can be used. Collection can be applied in various ways to equation (xxviii), the current version of LP transforms it to
$$3\cdot\cos(x) + \cos(x) = 0.$$
Now Factorization Preparation can be applied, producing
$$\cos(x)\cdot(3 + 1) = 0$$
which simplifies to $4\cdot\cos(x) = 0$. So, in this case Collection has been substituted for New in the schema.

LP applies Factorization to the last equation, stripping off the factor 4. LP then chooses which subschema to use for the remaining factor $\cos(x) = 0$. In fact, both subschemas contain identical methods, one application of Isolation. LP applies Isolation and solves the equation.

Note that LP does not really find the best method for solving this equation. The equation can be solved with two applications of Collection, followed by Isolation, or by Change of Unknown, followed by Polynomial Methods and Isolation.

4.5.4.4 Achieving the Purpose of a Section

Consider first the worked example shown in figure 4-7, which shows the solution of the equation
$$4\cdot\log_x 2 + \log_2 x = 5.$$
(Note that no new methods are taught in this example).

From this example, LP produces a schema, part of which is shown in figure 4-8.[44]

Now, suppose that LP is given the equation
$$4\cdot\log_2 x + 2\cdot\log_2 x = 12$$
to solve, and that it uses the schema above. The purpose of the first section of the schema is to allow Change of Unknown to apply. In this case, Change of Unknown can be applied to the given equation, substituting y for $\log_2 x$. There is no need for the Homogenization step. The program then proceeds to the next section of the schema, and solves the changed variable equation. In this case there is only one answer, $y = 2$, producing the substitution equation $\log_2 x = 2$. Now the program needs to choose which subsection to use to solve the substitution equation. Both sections are in fact the same, one application of Isolation is sufficient to solve the equation.

---

[44] In the schema, Homog stands for Homogenization and C.O.U. for Change of Unknown. The "identical-subterms(X,Eqn)" precondition holds if all the occurrences of the unknown X in Eqn are in identical subterms, see section 2.3.1.7. The Homogenization precondition "multiple-offenders(X,Eqn)" holds if there are multiple offending terms with respect to X in Eqn, and is-poly(X,Eqn) holds if Eqn is a polynomial in X.

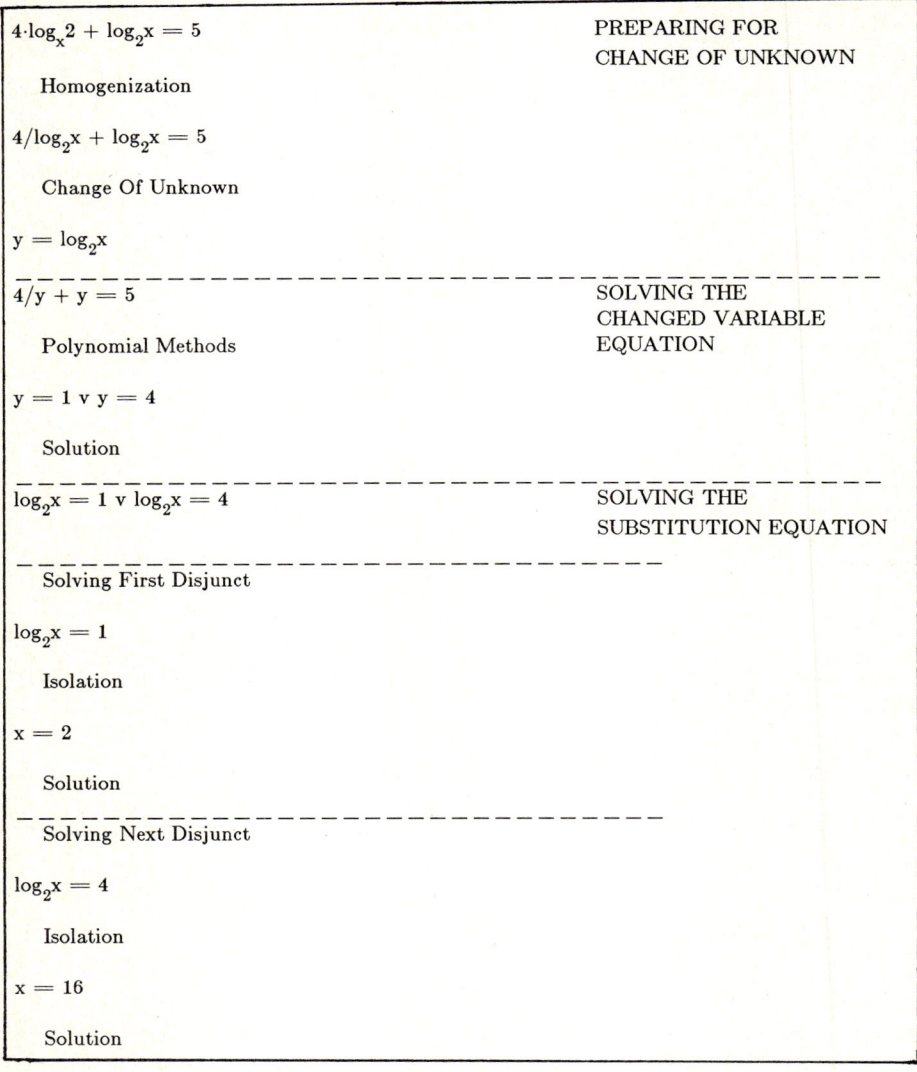

Figure 4-7: A New Worked Example

This example is similar to the one in the section above, in that LP does not find the best way to solve the equation.

4.5.5 Using the Solution Trace as a Worked Example

During the solution process, the given equation is successively transformed into new equations, until the solution state is reached. LP stores these successive states in a structure called the **solution trace**. The solution trace is a list of equations, each one following from the previous one by a legal move of algebra. In other words, the solution trace resembles a worked example!

NAME	SATISFIED PRECONDS	PURPOSE
Homog.	{mult-offenders(X,Eqn), mult-occ(X,Eqn)}	identical-subterms(X,Eqn1)
C.O.U.	identical-subterms(X,Eqn)	(Key Method)
Polynomial Methods	is-poly(X,Eqn)	(Solution)
Isolation	single-occ(X,Eqn)	(Solution)
Isolation	single-occ(X,Eqn)	(Solution)
Generating Equation: $4 \cdot \log_x 2 + \log_2 x = 5$. Unknown: x Schema Type: C.O.U.		

Figure 4-8: Part of the Schema

LP is able to use its solution trace as a new worked example. There is no point in doing this if it has solved a problem by following a schema exactly. In other cases, the problem may have been solved without a schema, perhaps after false starts, or LP may have used the schema, but with extensive modifications. In either of the latter cases, it may be helpful for LP to produce a new schema, so that similar problems can be solved more efficiently.

This ability of LP to treat its solution trace as a worked example gives it some of the flavour of **learning by doing**. Learning programs such as LEX, [55, 56, 54], and SAGE, [44, 45], adopt a similar approach. These programs use a problem-solving component to search part of the space, and the trace produced provides positive and negative instances for the learning component.

Training programs on solution traces has the disadvantage that the program may train on non-optimal solutions. This happens because if the space is interestingly large, the problem-solving component will only be able to search a small portion of the whole space. The solution found may then be non-optimal. It also has the disadvantage that the program will not be able to discover new operators, it just learns how to apply existing operators more effectively.

For these reasons, LP normally uses user-provided worked-examples. However, for LEX and SAGE, solution traces constitute the only type of example used by these programs, they are not given worked examples by a user.

4.6 Learning Existing Methods

LP can learn new methods. Can it also learn old ones, i.e. could LP start off with fewer methods, and learn some of the deleted ones?[45]

To some extent, the above examples show that LP can learn part of the PRESS Trigonometric Factorization methods, in particular LP can learn to solve equations that PRESS solves with the Arithmetic Progression Specialist (see section 2.3.1.9).

To answer this question in general, we must first consider how methods created by LP differ from the initial LP methods.

Firstly, LP usually creates a method that has just one rule on the rule list, whereas most of the initial LP methods contain several rules. However, if LP is given further examples, it can add more rules to the rule list, so this difference is unimportant.

A more important difference is that the methods created by LP tend to have less precise control information than the corresponding initial methods. A new method tries to apply its rules so that the postconditions of the method are satisfied, but there is no finer control than this. This contrasts with methods like Collection and Attraction, which only apply their rules to least-dominating terms (see chapter 2).

LP is able to learn some of the initial methods, whereas others prove more difficult. The following sections describe some successes and failures.

4.6.1 The Successes

4.6.1.1 Learning Collection

LP is able to learn a version of Collection (and Attraction) without learning the least-dominating condition. The resulting methods are just slightly more inefficient than the corresponding initial methods. Actually, LP learns more than one method that corresponds to Collection. Suppose that Collection is disabled[46] and that LP is given the following worked example.

$$3 \cdot \sin(x) + 4 \cdot \sin(x) = 0 \qquad \text{(xxix)}$$

$$7 \cdot \sin(x) = 0 \qquad \text{(xxx)}$$

$$x = 180 \cdot n_1.$$

LP cannot discover which operator transformed line (xxix) to line (xxx), and the user provides the rule

$$M \cdot Y + N \cdot Y \Longrightarrow (M + N) \cdot Y. \qquad \text{(xxxi)}$$

LP creates a new operator that applies this rule, whose purpose it to reduce the number of occurrences of the unknown to one so that Isolation can be applied. This is a common use of Collection, reducing the number of occurrences from two to one.[47]

This combination of applying Collection, followed by Isolation, accounts for a number of elementary techniques. For example, LP can learn how to solve quadratic equations, beginning with only

---

[45] Rather than rewrite LP so that it contains fewer methods, we simply use the disabling mechanism, which allows the user to prevent LP from using the specified methods. Effectively, LP then knows nothing about the methods that have been disabled.

[46] i.e. LP is not allowed to use either the Collection method or any of the Collection rewrite rules.

[47] The IMPRESS project, [16, 75], considered making this restriction of Collection a new method, called Whittle.

Isolation, and learning Collection as above.

Other examples use Collection in a different way, e.g.

$$\sin(4 \cdot x) + 2 \cdot \sin^2(2 \cdot x + 2 \cdot x) + \sin^3(4 \cdot x) = 1 \qquad \text{(xxxii)}$$

$$\sin(4 \cdot x) + 2 \cdot \sin^2(4 \cdot x) + \sin^3(4 \cdot x) = 1 \qquad \text{(xxxiii)}$$

$$y = \sin(4 \cdot x)$$

.
.
.

etc

Here Collection is applied so that Change of Unknown can be applied. Once the user has given the rule (xxxi), LP classifies this as a method that prepares for Change of Unknown. It is similar in this respect to Homogenization![48]

These are not the only examples, Collection can be used to prepare for other methods. LP learns each kind of application of Collection as a new method. This is useful given the way that LP plans, the new methods are suitably specialized. It has the disadvantage that certain rules, such as (xxxi), need to integrated into each of the new methods.

Also, as mentioned above, the new methods are slightly less efficient than standard Collection, as the rewriting is not restricted to least-dominating terms.

4.6.1.2 Learning Attraction

The situation with Attraction is very similar, except that the most common use of Attraction, to prepare for Collection, satisfies no new preconditions, as described in section 4.4.2.5. This creates no difficulty.

4.6.1.3 Learning Factorization Preparation

LP can learn Factorization Preparation, except that the method learned is more general. There is no need for the left-hand side to have common subterms, all that matters is that Factorization can be applied. For example, the step from (xxxiv) to (xxxv)·

$$\sin(x) + \sin(3 \cdot x) = 0 \qquad \text{(xxxiv)}$$

$$2 \cdot \sin(2 \cdot x) \cdot \cos(x) = 0 \qquad \text{(xxxv)}$$

can be done by the method learned by LP. (PRESS solves this using the trigonometric specialist methods described in appendix I.3.

The situation here is the reverse of that of Collection. The initial Factorization Preparation method has the extra common subterms preconditions to allow greater planning. The new method learned by LP loses this, but it can be used in more situations.

---

[48]Note that Homogenization does not apply directly to equation (xxxii), as the assumption about the form of angles is violated, see section 3.1.3.5.

4.6.1.4 The Logarithmic Method

LP learns the Logarithmic Method, described in section I.4, but it does not learn one of the unimportant details, which base of log to use. Nevertheless, the method still works.

4.6.2 Learning the Key Methods

No attempt has been made to make LP learn the key methods, due to the structure of the current code. The key methods play too central a role for them to be removed. For example, the current implementation contains a "black-box" polynomial package, and this is used by many parts of the code. However, LP can certainly learn new rules for these methods, an example for Isolation is shown in appendix V (see also section 4.9.5).

There should be no difficulty in principle in learning the complete Isolation method, it satisfies the missing precondition of the goal state that the equation is solved. The current implementation does not use this piece of information, because it already assumes that Isolation is a key method that produces a solution!

The next implementation of LP should allow at least Isolation to be learned, if not the other key methods.

4.6.3 Less Successful: Learning Homogenization

Some initial methods make extensive use of control information, Homogenization, for example. As described in chapter 3, Homogenization has a set of rules for choosing the reduced term for the various types of offenders set. This method can be viewed as a set of submethods, each working on one type of equation. These submethods produce the same postcondition, Change of Unknown can be applied to the new equation.

This should not cause problems, it seems to be not very different from the Collection case described above, where the submethods are differentiated by what method is applied next, rather than the type of equation. However, it is difficult to learn even one of these submethods entirely. Homogenization uses just too many rules, applied in complex ways. As described in chapter 3, many of these rules can loop. To learn which rules should be applied, and when, it seems that LP would need several examples and some sort of concept-learning.

At present, LP is able to learn Homogenization rules from examples where only one rule needs to be applied, see section 4.9.5. If several rules need to be used, one after the other, LP can get bogged down as it tries to apply the rules in various ways.

In summary, the present version of LP cannot learn the full Homogenization method. To do so seems to require concept-learning techniques.

4.6.4 Conclusions

In summary, some of the existing methods are probably too hard for LP to learn without modification, full Homogenization being an example. Learning even parts of these methods may prove too difficult. Other methods, such as Collection are much more tractable. LP does not actually learn methods that correspond exactly with the initial methods, sometimes it learns several methods in place of one, sometimes it learns a more general method. Also, the initial methods have been optimized and LP creates methods that are more inefficient.

4.7 Possible Extensions

LP can be extended in a number of ways. The following sections describe some of the possibilities.

4.7.1 Adding Concept Learning

LP mainly learns from one example. While it can do some useful things if it is given more than one example, such as adding an extra rule to a new method, LP fails to generalize from additional examples. This situation could be improved if LP incorporated some sort of concept-learning.

4.7.1.1 Concept Learning Programs

A concept-learning program has to learn a symbolic description that enables it to determine whether or not an "object" is an instance of the target concept. The "object" is usually a symbolic description of a real object, but in theory one could imagine presenting a real-world object to a visual sensor, the computer then produces the description. We call the object (or symbolic description) that is currently being given to the program the **current training instance**. The current training instance can either be an example or non-example of the target concept.

Perhaps the most famous example of a concept-learning program is that of Winston, [88], that can learn the concept of an arch. The Focussing algorithm of Young et al, [19, 20, 89] is an extension of Winston's.

4.7.1.2 Precondition Analysis and Empirical Concept Learning

What relevance has concept-learning for Precondition Analysis?

In [20], we surveyed several rule-learning programs. Each of the programs has two main parts: a critic for identifying faulty rules and a modifier for correcting them. Somewhat surprisingly, most of the modification techniques used by the rule learning programs are subsumed by the Focussing algorithm, see [20] for details.

While LP is not a rule learning program, (it learns control information), it seems that some of these techniques may be applicable. Mitchell has described how standard concept-learning methods (Mitchell calls these **empirical** methods) can be mixed with more analytic methods, [53]. His LEX2 program, [56, 54], which does this, is described in the section 5.4. However, some of Mitchell's ideas are not directly applicable to LP (see section 5.4), so here we consider how empirical learning techniques could be used by LP.

Concept-learning can be used by LP in different ways. One way is to allow LP to produce a schema as it does at present, but then use concept-learning to allow it to discover the class of equations that the schema solved, the **target-class** of the schema. We could use the Focussing algorithm to do this. We could just tell the program that certain equations were examples and non-examples, as in Winston, [88], or would could let LP attempt to use the schema, and discover this information itself, like LEX, [55]. In either case, the algorithm may begin to converge, thus producing a partial description of the required class. This could be done for a set of schemas. Then, when given a new equation, LP would classify it and use the descriptions produced by the focussing algorithm to decide which schema should be used. This is somewhat similar to the original version of LEX. LEX has to use partial information to choose an operator, in this case LP uses partial information to choose a schema.

Such an approach allows the empirical methods to supplement LP's analytical techniques. Once the schema method had been chosen, LP would operate exactly as at present. For example, if the target class had not been precisely identified, the schema would still need patching, and the techniques used at present would still be valid. The role of the empirical methods in such a program is to guide the selection of the schema. Once this is done, LP would execute it in the same way as it does now.

Note that the above process can be done in various ways. We could allow LP to ignore the preconditions of schema methods, and use a whole new classification produced by the concept-learning process. Alternatively, and more satisfactorily, we could allow concept-learning to modify the preconditions of the schema methods.

Suppose that LP has created a schema method from a worked example and is given an equation to solve. Assume that LP tries to use the schema and that the given equation is not identical to the generating equation of the schema.

If the attempt fails, the preconditions of the schema method could be made more specific. One way of doing this is have a set of standard conditions, and find which ones that are satisfied by the positive instance (the generating equation) are not satisfied by the negative instance (the given equation). Some subset of these conditions would be added to the preconditions of the schema method. This is a common concept-learning approach, and seems a reasonable first try. One disadvantage is that there is no new analytic input, the added preconditions do not fit in with the previous analysis of the example. One more satisfactory approach involves arranging the conditions in sets of hierarchies, as in the focussing algorithm, see [20, 89]. Each hierarchy contains related conditions, arranged in order of generality, e.g. a possible hierarchy for the number of occurrences of the unknown is

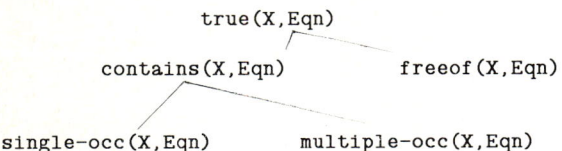

The root relation is always true, conditions nearer the tips are more specific than those nearer the root. So, one way of making the preconditions of the schema method more specific is to replace certain conditions by those nearer the tips in the hierarchy. Thus if the generating equation contains 2 occurrences of the unknown, the schema method initially has as one precondition **contains(X,Eqn)** (unless the schema analysis indicates a stronger condition). If the schema fails on a given equation with one occurrence of the unknown, the precondition is strengthened to **multiple-occ(X,Eqn)**.

If a schema method succeeds on a given equation, the preconditions can be generalized by replacing conditions with those nearer the root.

Note that none of these methods is certain to work well. There may be situations in which the program gets "confused", when it is given one example that causes the conditions to be generalized, while the following negative example requires the conditions to be made more specific again. Such situations arise because of the nature of the domain, preconditions are necessary not sufficient. For this reason, specification is probably safer than generalization, as the preconditions are too "optimistic" anyway.

The preconditions of the new ordinary methods can be treated in a similar way. New examples provide new instances of the new methods, and when LP tries to solve equations, it will sometimes succeed and sometimes fail in its attempts to apply a new method. These positive and negative instances allow the program to use empirical concept-learning techniques on the preconditions of the new methods. Note that this variation may affect the schema methods as well, as the latter take their preconditions from methods.

Section 6.3.3 further discusses the possibility of adding concept-learning to Precondition Analysis.

Learning Control Information

As discussed in section 4.6, some of the initial methods have complex control information apart from the preconditions and postconditions that LP currently cannot learn. A concept-learning program could help here, learning which trigonometric term to rewrite in Homogenization for example. Such a concept-learner could be rather separate from the rest of the program, as there is little need for integration. Alan Bundy has done some experiments here, teaching the focussing algorithm conditions for Isolation, [12, 11].

Summary

This section has described one way in which concept-learning can be added to LP. The benefits are not certain, but the author plans to implement a version of LP that includes these techniques.

Mitchell's analytic concept-learning method may also prove useful, see section 5.4.

4.7.2 Extending the Use of Schemas

Another extension concerns a more flexible use of schemas. At present, LP chooses a schema method, and uses it until the equation is solved, or the schema fails. In the latter case, it tries to use another schema method from the beginning. If all the available schema-methods fail, LP tries to solve the equation without any schema. LP tries to use each schema method from the top, this is somewhat inflexible.

An alternative approach is to allow LP to use any part of any schema at any time. LP could start on any line of any schema whose preconditions were satisfied. If this schema eventually fails[49], instead of starting again, LP would iterate the process for the new state of the equation, i.e. it would again find a line in a schema whose preconditions were satisfied by the current equation, and begin the schema from that point.[50]

Such an approach obviously increases flexibility, but at some cost. It can cause a mini combinatorial explosion in the meta-level search space.[51]

To reduce the potential combinatorial explosion, this extension should perhaps be limited to the case where the current schema has failed after all attempts at modification. Instead of returning to the original given equation, LP finds a step in a schema whose preconditions are satisfied by the current state of the equation.

Some additional tests could be carried out, such as the extra functional type test used by LP at present (see section 4.5.1). If such a schema can be found, it can be used without too much danger,

---

[49] Here, there are several possible meanings of failure. At one extreme, we could say that the schema fails if the step listed in the schema cannot be applied at once. At the other end, the program may try all possible schema modifications before the schema fails.

[50] This is similar to PLANEX, see section 5.3.

[51] If, additionally, each attempt to use part of a schema can use all the schema modification processes used by the current version of LP, the explosion is even worse.

the schema was used to solve a fairly similar problem. This overcomes the objections of section 4.5.3.1, the new schema overrides the waterfall. However, the schema must be found at once, no transformations are permitted first, or these criticisms described will apply.

The above extensions allow LP more flexibility in the use of the schemas. This increase of flexibility is obtained at the cost of extra search at the meta-level. This is because the above modifications introduce choices where there are no choices now. For example, at present, when all the schema methods fail LP returns to the given equation and behaves like PRESS. With the extensions, it would still have the option of doing this, but alternatively it could look for a new schema for the current equation, or try to apply a step then look again etc.

There is no easy way to decide whether the increase in flexibility outweighs the extra cost, it really depends on which problems are to be solved.

4.7.2.1 Summary

Two modifications have been suggested in the previous sections: by changing the way that schema methods are used, or adding a concept learning component. There is really a spectrum of changes, trading flexibility against increased search.

However, these involve fairly minor changes to LP.

A more major change involves applying Precondition Analysis to a domain other than equation solving. Which domains are suitable, i.e. how generally applicable is Precondition Analysis? We will now discuss this question.

4.8 The Applicability of Precondition Analysis

Unsurprisingly, Precondition Analysis is not applicable to all domains.

Recalling McCarthy's advice, [50]:
> "In order for a program to be capable of learning something, it must first be capable of being told it."

we consider the problem solving element. Precondition Analysis learns control and strategic information, this information is used by the problem-solver. Therefore, the technique is only useful for domains where the problem-solver has a large space to search, and the search must be constrained. This is true of most interesting domains!

The control information specifies when and how operators should be applied. In the case of LP, these operators are methods. In general, operators are used to transform the current state into another. For LP, the only objects of interest are the equations. More generally, operators can transform several objects of concern. So, the domain must admit such operators. This is not very restrictive. Many of the commonly used learning domains allow operators, and the relevant learning programs are operator based, e.g. LEX and symbolic integration ([55, 56, 54]), SAGE and Towers of Hanoi ([44, 45]), ALEX and ELM solving linear equations ([58, 8]), and STRIPS and robot task planning ([35, 36]).

The operators must also "fit" together in some way. The sequence of operator applications used in the worked example is not any old sequence, each operator application serves some purpose. This should also be a feature of most domains.

An important restriction is that the control information for each operator must be explicit, i.e. the meta-level information must be separate from the object-level. Although most domain operators can be structured in this way, many current learning programs do not have a clear distinction between object and meta-level. Of course, programs using meta-level inference meet this requirement.

The program must also be given a lot of knowledge about the domain. For example, the program must be given (among other things) the following:

- a set of operators,

- a set of termination conditions (i.e. conditions that detect when the problem is solved),

- a set of possible preconditions,[52]

- which syntactic features to concentrate on,

- what operators are key operators[53], and

- how to detect when the linear arrangement of the worked example hides non-linear steps. (See section 4.3.1).

Therefore, Precondition Analysis cannot be used on domains that are not well-understood. (The same is true of most other learning techniques *in practice*).

Note, however, that although the program does not require a large amount of knowledge per se, it must know a lot at the level of the task. This means that Precondition Analysis is applicable to a wide range of ability. For example, a version of LP, knowing only Isolation, can learn a form of Collection, see section 4.6.

In conclusion, Precondition Analysis can be applied to many domains, providing that certain restrictions are met. The operators of domain must have the object and meta-level components separated (as in meta-level inference), and the program must be given a lot of knowledge about the domain.

The **Explanatory Schema Acquisition** technique of DeJong [30, 31], has many features in common with Precondition Analysis. His program is given newspaper-type stories concerning events such as a kidnapping incident, and an item about a student damaging the brakes of his teacher's car because he was given a low grade. The program explains events in terms of satisfying preconditions of subsequent events, this is close to the explanation produced by LP.

However, DeJong is concerned with knowledge-based generalization rather than problem-solving. His program builds new concepts from the examples, e.g. kidnapping is seen as a novel way of combining stealing with bargaining. These concepts are then used to understand later stories, but there is no sense in which they are executed. Nevertheless, DeJong's work shows that it is possible to analyze preconditions to produce useful results in domains very different from equation solving.

---

[52]LP is given this implicitly, the only preconditions it uses are those in the existing operators. It may be possible to extend this.

[53]This information is not essential, there may be no key operators, see section 4.3.2.3.

4.9 Evaluation and Results

The first part of this section describes some tests performed to evaluate the learning potential of LP. The remainder shows examples of LP learning new and existing methods.

4.9.1 Evaluation

How well does LP learn? There are several dimensions along which the performance of LP can be evaluated. Obviously, LP should be able to solve more equations as the learning process proceeds. However, it would also be desirable for LP to solve equations faster as it learns. Unfortunately, these two requirements are often in conflict.

4.9.1.1 The Problem

There are two related reasons for this. Firstly, the schema execution mechanism introduces additional overheads. Processes that introduce these overheads include:

- Checking if any schemas exist.

- Choosing between the possible schemas.

- All of the patching mechanisms.

Generally, LP will attempt to use a schema if one is available. In some cases, this may cause some problems. Consider the following sequence:

1. LP is initially given an equation (E say) that it can solve efficiently without using any schemas.

2. LP is then given its solution trace to process as a worked example. This allows to build a schema for E.

3. LP is given E to solve again.

If E is very simple, the overheads mentioned above will cause LP to solve the equation rather slower than it did originally, giving the impression that learning has degraded the LP's performance. This only happens if the original solution required very little search (e.g. the solution used only Isolation, a method high in the default priority ordering.) If the equation is slightly more complex, the solution time at stage 3 is usually less than the original solution time. In this type of case, the reduction in search time more than makes up for the schema overheads.

More realistically, the same sort of problem may arise if LP is given a worked example for a different equation at stage 2. The schema may require extensive patching at stage 3, again leading to an increased solution time.

The other reason why the two requirements above conflict is that as LP learns more schemas and methods, it obviously has more choices at each stage. For any particular equation, many of these choices are not helpful, but they all need to be considered. If LP has learnt several new schemas, it may have to try them all before finally trying successfully to solve without a schema. Similarly, new non-schema methods can increase solution times, as LP has to consider using them before trying some of the older methods. Increasing the number of methods also increases the time taken during schema patching, as again more methods have to be considered.

4.9.2 The Terms of Evaluation

The above discussion has shown that we cannot expect LP to both solve more equations and to solve them faster. LP was designed with the former goal in mind, so we will discuss LP's performance primarily in terms of number of equations solved. However, we will also mention solution time.

4.9.3 Tests and Results

The first set of tests examined LP learning from its initial state. Recall that LP is already fairly expert at this level. The second set tested how LP well could be built up from knowing only Isolation.

Both sets used an A level test set, containing 104 questions.

4.9.3.1 LP as an Expert

Initially, LP could solve 88% of the test set problems. The first test involved giving LP a worked example for the equation
$4 \cdot \cos(x) + 3 \cdot \sin(x) = 1$.
The worked example teaches LP the Collection rule:

$A \cdot \cos(X) + B \cdot \sin(X) \Longrightarrow R \cdot \sin(X + \tan^{-1}(B/A))$
where
$R = (A^2 + B^2)^{(1/2)}$.

LP can solve equations of this type without the rule, using Homogenization, but the solution process is rather lengthy, involving Homogenization, Change of Unknown, Nasty Function Method, Polynomial Methods, and Isolation. After the example, LP was given the test set again. Although LP could not solve any more of the equations, all the equations of the above type were solved significantly faster. For example, the equation
$2 \cdot \sin(x) + \cos(x) = 1$ (AEB 1972)
took LP 4.3 seconds initially, and only 1.1 seconds after the worked example.[54] Initially, the average time spent on an example was 5.7 secs. On the second run this was reduced to 4.3 secs. (In each case four problems exceeded the allowed time limit.) So, evaluating on timing grounds, the overall performance of LP was increased by the new example. Analysing the successes and failures separately, the average time taken on successes decreased from 5.6 seconds to 4.3 seconds. However, the average time for failures increased from 2.7 seconds to 4.1 seconds.

Next, LP was given the worked example for the equation
$\cos(x) + 3 \cdot \cos(2 \cdot x) + \cos(3 \cdot x) = 0$,
a slight variant of the running example of this chapter. This example teaches LP a new method, and the rule for the sums of cosines. (In this case, LP couldn't solve the equation without the example.) Additionally, LP was also given the rules for the difference of two cosines, and for the sums and differences of sines.[55] LP was then able to solve 6 more of the problems, because of the new rules, the

---

[54] All timing information for LP running on a Dec-10 KL. Timing commences when LP is given the equation, and finishes when the equation is solved or when LP reports that it cannot solve the equation. On some problems, LP did not finish within five minutes, such runs were stopped. These problems were counted as failures. The timing information does not include such cases, which are indicated separately.

[55] These rules were given to LP 'directly', LP merely checking that the rules are suitable. We could have given LP a new example for each of the rules, but this is too time consuming.

new method and the schema. However, the timing figures were rather worse; the average time rising from 4.3 seconds (for the initial run) to 6.3 seconds, now longer than the original average time. Successes now took an average of 5.6 seconds, failures 23 seconds. (Two problems exceeded the five minute time limit) The large size of latter figure is due to the effect described above; as LP learns more methods and schemas, it has more things to try before eventually failing.

The third test added the example for the equation.
$$\cos(4\cdot x) + \cos(6\cdot x) + \sin(4\cdot x) + \sin(6\cdot x) = 0 \qquad \text{(AEB 1973)}$$
This example increased to the number of problems solved by only 2, (but only 4 problems remained unsolved). The timing figures are more complex. Generally, the timing benefits experienced in test 1 and 2 still remained; by using the correct schemas, LP was able to solve equations of the form
$$A\cdot\cos(X) + B\cdot\sin(X) = C$$
as almost as quickly as it did in the first test, and similarly for problems with sines or cosines in arithmetic progression. However, LP was rather slower on problems where the schemas were of no use, as often LP tried to use at least one of the schemas. The average time taken was 10.1 seconds, averaging 9.5 seconds on successes and 32 seconds on failures. (One problem exceeded the time limit).

This picture continued for the next test, which gave an example for the equation
$$\sin(2\cdot x) + \sin(3\cdot x) + \sin(5\cdot x) = 0$$
This allowed LP to solve 1 more problem, (now 98%), with average time up to 12.5 secs; 17.1 secs for successes and 44 secs for failures. (No problems exceeded the time limit.) The results of the tests are summarized in table 4-2.

Test	No. of new schemas	No. of new methods	No. of new rules	No. solved problems	Time Successes	Time Failure	Time All
1	0	0	0	92	5.5	2.7	5.7
2	1	0	4	92	4.3	4.1	4.4
3	2	1	8	98	5.5	23.5	6.3
4	3	2	9	100	9.5	32.8	10.2
5	4	3	10	101	16.3	44.1	12.4

Table 4-2: Results of Evaluation

4.9.3.2 LP as a Novice

For this set of tests, all the LP methods were disabled except for Isolation. Initially, LP could solve only 4 of the problems. The average time taken for each problem was very low, 0.4 seconds, but this is not surprising, as there were so few possibilities. LP was then given a worked example for
$$4\cdot\cos(x) + 3\cdot\sin(x) = 1.$$

This example taught LP a method that prepares for Isolation, i.e. a restricted version of Collection. After the example, LP could solve 36 of the problems. The average time was 1.7 seconds.

Next, LP was given the same example, but it was also given several rules for the new method. This allowed LP to use all the standard Collection rules for its new method. (This doesn't make the new method the same as Collection, as it can only be applied to prepare for Isolation). This allowed LP to solve one more problem, but the average time increased to 3.8 seconds.

Teaching LP from this low level is obviously rather time-consuming, and no more tests were performed for this level. Generally, more than one or two methods are required to solve problems of this difficulty. Nevertheless, the results of the second set of tests shows (to some extent) the same kind of learning behaviour of in the first set.

4.9.4 Conclusions

These tests show that LP is able to improve its performance, even when starting at a rather high level. By the conclusion of the first set of tests, LP was able to solve 9 more problems, having learnt 10 new rules, 3 new ordinary methods and 4 schema methods. Unfortunately, the increase in equation solving power was at the expense of increasing solution times.

4.9.5 Learning the Initial Methods - Examples

As described in section 4.6, LP is capable of learning (to a useful extent) a number of its initial methods. It has of course also learned a number of new methods, mainly in the field of Trigonometric Factorization. This section gives a few examples of LP learning some existing methods.

4.9.5.1 Learning the Attraction Method

If LP is given the example

$$\log_e(x + 1) + \log_e(x - 1) = 3 \qquad \text{(xxxvi)}$$

$$\log_e((x + 1) \cdot (x - 1)) = 3 \qquad \text{(xxxvii)}$$

$$\log_e(x^2 - 1) = 3 \qquad \text{(xxxviii)}$$

etc

with Attraction disabled, it will not recognize the step from (xxxvi) to (xxxvii). The following step is Collection, but no new preconditions are satisfied by the Attraction step. LP creates a method that prepares for Collection in some unspecified way, applying the (now user-supplied) rule

$$\log_A X + \log_A Y \implies \log_A X \cdot Y.$$

This method can now be used on other similar equations, such as

$$\log_e(2 \cdot x + 1) + \log_e x + \log_e(x - 1) = 2$$

4.9.5.2 Learning New Isolation Rules

It is possible to teach LP about new functions. If LP is given the (very short) example
$$f(x) = 1$$

$$x = g(1),$$

it first warns the user that f and g are unknown functions. It then recognises the step as a new type of Isolation, and prompts the user for a rule. If the user supplies the rule

$$f(A) = B \quad \rightarrow \quad A = g(B),$$

LP stores this as a new Isolation rule, and adds f and g to its list of known functions. It can now solve examples of varying complexity, e.g.

$$(f(x))^2 - 2 \cdot f(x) + 1 = 0$$

solving this by Change of Unknown, Polynomial Methods and Isolation, to arrive at the answer $x = g(1)$. This example is shown in full in appendix V.

4.9.5.3 Preparing for Change Of Unknown

As remarked in section 4.6, LP is unable to learn Homogenization fully. However, given the worked example of figure 4-7 above, with Homogenization disabled, LP is able to learn that the new rule
$$\log_A B \implies 1/\log_B A$$
can be applied to allow Change of Unknown to be applied. Thus, it can solve equations such as
$$2 \cdot \log(2, x) + \log(x, 2) + 2 \cdot \log(2, x) = 4.$$
Here, LP tries various ways of applying the new method, but it discovers the correct one, (rewriting the second log term), as this is the only one that satisfies the preconditions of Change of Unknown.

Many similar Homogenization rules have been taught.

4.9.6 Trigonometric Factorization Examples

This section contains examples of methods learned by LP that are not included in LP initially. Some of these methods are in the PRESS Trigonometric Factorization methods, but LP is able to use these methods to solve equations that PRESS currently cannot solve.

The example used throughout this chapter, the solution of
$$\cos(x) + 2 \cdot \cos(2 \cdot x) + \cos(3 \cdot x) = 0$$
shows how LP learns to add two cosine terms to produce common subterms for Factorization Preparation.

LP has learned other ways of transforming an equation so that Factorization can apply. For example, the equation
$$\sin(x) + \sin(3 \cdot x) = 0$$
is solved by adding the two terms to obtain
$$2 \cdot \sin(2 \cdot x) \cdot \cos(x) = 0,$$
which is then factorized. In this case, adding the two terms does not produce common subterms, but Factorization can be applied directly.

LP has also learned another way of creating common subterms for Factorization Preparation, applying the rule
$$\sin(2 \cdot X) \implies 2 \cdot \sin(X) \cdot \cos(X),$$
to equations such as
$$\sin(2 \cdot x) = \cos(x).$$
These two methods allow LP to solve the equation
$$\sin(2 \cdot x) + \sin(3 \cdot x) + \sin(5 \cdot x) = 0,$$
that PRESS cannot solve.

4.9.7 Using Hints

Another example that LP solves that PRESS cannot is:
$$(\operatorname{cosec}(2 \cdot x) + \cot(2 \cdot x))^2 = \sec(2 \cdot x). \qquad \text{(A.E.B. 1973)}$$
This question comes with a hint. The student is first asked to show that
$$(\operatorname{cosec}(x) + \cot(x))^2 = (1 + \cos(x))/(1 - \cos(x)).$$

Several exam questions do supply hints. They often serve to focus the student's attention on a particular identity, as in this case. The identity is used to solve the second part of the question, solving an equation. Sometimes the identity will be new to the student, this is presumably true in this case. At present, PRESS cannot use such hints.[56] By giving suitable examples, LP *can* be taught them.

LP is taught the hint on a worked example. This example is contrived, but it is one way of teaching LP the hint, without greatly restricting its applicability.

$x + (\text{cosec}(x) + \cot(x))^2 - (1 + \cos(x))/(1 - \cos(x)) = 2$

$x + (1 + \cos(x))/(1 - \cos(x)) - (1 + \cos(x))/(1 - \cos(x)) = 2$

$x = 2$.

The second step is a Collection step (which also solves the equation). The first step does not satisfy any missing preconditions, so, once LP has been given the identity, it creates a method that has as preconditions and postconditions **multiple-occ(X,Eqn)**, with a suggestion that Collection is applied next. LP also creates a schema method from the example.

When given the new equation, LP tries to use the schema method. The first step in the schema method, applying the new method, succeeds, producing

$(1 + \cos(2 \cdot x))/(1 - \cos(2 \cdot x)) = \sec(2 \cdot x)$

but the attempt to apply Collection fails. The purpose of applying Collection is to solve the equation, so LP tries to substitute for this step. None of the methods known to be capable of doing this, Isolation or Polynomial Methods, can be used, so LP tries to add another step that maintains the currently satisfied precondition (multiple-occ(X,Eqn)). Two methods are applicable. Firstly, Change of Unknown could apply, performing an uninteresting substitution, $y = 2 \cdot x$, but this is not allowed as this method creates extra subgoals. Homogenization is used, performs the rewrite $\sec(2 \cdot x) \Longrightarrow 1/\cos(2 \cdot x)$. The equation is now

$(1 + \cos(2 \cdot x))/(1 - \cos(2 \cdot x)) = 1/\cos(2 \cdot x)$.

In fact, Change of Unknown should now be used, substituting y for $\cos(2 \cdot x)$, but this is not allowed, and LP tries to apply Collection again. This fails and finally the schema fails. At present, LP returns to the original equation, and performs the *same* steps as above, possibly (depending on the method ordering) including the extra substitution. Now, however, the Change of Unknown step can be used, and after an application of Nasty Function Methods, the equation becomes

$y \cdot (1 + y) = (1 - y)$.

LP successfully solves the equation, using Polynomial Methods, and solves the substitution equation using Isolation.

In this example, it would obviously be better if LP did not return to the given equation if the schema fails, LP just has to duplicate the work later on. Nonetheless, it does eventually solve the equation, and the worked example used is not typical, so it is wrong to generalize on the basis of this example.

Why does PRESS fail? As PRESS cannot use the hint, it applies Homogenization straight away, rewriting cosec and cot in terms of cos. This leads to a fairly large rational expression after Change of Unknown. The Nasty Function Methods multiply the equation through to clear the denominator,

---

[56]The author has written some code that allows PRESS to make use of such identities, but it is rather restricted. This code is not included in the current version of PRESS.

and square to remove square roots. The resulting equation is very large, and PRESS gets bogged down. Although the expression is actually correct, PRESS runs out of space manipulating it.

4.10 Conclusions

This chapter has presented the learning technique of Precondition Analysis. This technique has been implemented in the equation solving program LP. Precondition Analysis learns control information in the form of new operators and learns strategies in the form of schemas. This knowledge can be obtained from a single example.

A program using Precondition Analysis works in two phases, the learning cycle and the performance phase. In the learning cycle, the program is given an example of a correctly executed task. The program builds an explanation of the strategic reasons for each step of the task. This explanation is in terms of satisfying the preconditions of following steps. From this explanation, it builds a plan that is used in the performance phase. The schema is executed in a flexible way, using the explanation to guide the problem solving attempt.

Precondition Analysis relies on having control information explicitly represented, programs using meta-level inference have this property.

This technique is knowledge-based, the program must have a lot of knowledge about the domain at the level of the task.

The LP program

Precondition Analysis has been successfully implemented in the LP program. LP is a version of PRESS that learns new methods and constructs equation-solving plans. It uses Precondition Analysis to learn from worked examples.

LP has successfully learned a number of new methods and plans. Examples can be seen in appendix V.

4.10.1 Precondition Analysis and Meta-level Inference

In section 1.2, it was claimed that one advantage of meta-level inference is that it enables the learning of strategic information, as well as factual knowledge.

The technique of Precondition Analysis supports this claim, LP learns strategic knowledge, taking advantage of the fact that the control information is separate from the factual knowledge.

Precondition Analysis relies on the separation of factual and control knowledge, this separation occurs in programs using meta-level inference. Therefore, if Precondition Analysis is to be used, the underlying problem-solving program should use meta-level inference. Meta-level inference has other advantages, these were listed in section 1.2.

In our view, a program using Precondition Analysis with meta-level inference is capable of learning powerful strategies in domains with a large search space. LP is an example of such a program.

There are, of course, other techniques for learning control information. Some of these are discussed in the following chapter.

CHAPTER 5

RELATED WORK

This chapter describes some of the work of other authors that is related to the material in this book. Such work can be divided into four areas:

- Algebraic manipulation,

- meta-level inference,

- learning, and

- planning.

Algebraic manipulation has already been discussed briefly, MACSYMA was compared with PRESS in section 2.6. The remaining areas obviously form a large part of A.I. and thus cannot be discussed fully. This chapter is structured as follows:

1. Section 5.1 describes some other programs that make use of meta-level knowledge.

2. ALEX, [58] is described in section 5.2. ALEX also learns equation solving procedures by examining worked examples, but differs from LP in many ways.

3. Section 5.3 describes STRIPS and its derivatives. STRIPS can be viewed as both a planning program and a learning program, and is described in some detail.

4. Section 5.4 describes the LEX2 program of Mitchell et al. This program learns heuristics for symbolic integration, and uses a technique somewhat related to Precondition Analysis. Utgoff's extension of Mitchell's method is also described in this section.

5. The chapter concludes with a summary.

These programs have been singled out for discussion because they appear to be more closely related to the work described in this book than most of the other work in the areas of interest.

5.1 The Use of Meta-Knowledge

Recently, many programs have been built that utilize meta-knowledge. A program that uses meta-knowledge has two[1] levels, an object-level and a meta-level. Such programs include those

---

[1] Or more. Stefik's MOLGEN, [73], has three levels.

described in this book, but most of the uses of meta-knowledge are quite distinct from meta-level inference. To illustrate this, we examine the programs of Hayes, Weyhrauch, Davis & Buchanan and Lenat. These programs are obviously a small sample, but together they demonstrate some of the various differences.

5.1.1 GOLUX

The idea of using logic to control search is not new; Hayes used it in his GOLUX project, [37]. However, our approach differs from that of Hayes. PRESS and LP use a control language that is domain-specific, the control information is expressed in the meta-theory of algebra. In contrast, the control language of GOLUX is general-purpose, but it is very limited in scope, it merely specifies which of the different kinds of resolution restrictions should be used. The greater specialization of the control language of PRESS and LP allows the programs to make far more directed searches than GOLUX. The use of meta-level inference allows our programs to select which object-level rule should be applied next, GOLUX could only vet the result after the rules had been applied, in order to make sure that their applications met the restrictions imposed by the control language.[2]

5.1.2 FOL

FOL, [84, 1, 85], is able to prove meta-level theorems, and it can use these theorems in object-level proofs by means of Weyhrauch's "reflection principles", [84]. The meta-level can use information contained in the object-level, and knowledge possessed by the meta-level knowledge can be reflected into the object-level. However, FOL does not use meta-level inference to guide the search.

In [1], FOL is applied (interactively) to the domain of algebra. Weyhrauch shows how the use of meta-theoretic knowledge improves the quality of the (object-level) proofs, in that they are both easier to find and easier to understand.

In [85], Weyhrauch shows how FOL can introduce derived inference rules to the object-level theory. FOL is given a formalized reasoning system. Some of the inference rules are not in a usable form, but by using the theorem proving/checking facilities of FOL, the user can transform these rules to more useful forms.[3] Such transformations involve the reflection principles.

Weyhrauch points out that most theorem provers have fixed formalizations. These formalizations state the syntactic aspects of what it means to be a formula, and the inference rules of the system. FOL, however, is much more flexible, because of its ability to perform transformations.

Search does not appear to be a problem for FOL, presumably because the system is so interactive. We use meta-level inference to constrain search, Weyhrauch uses meta-level knowledge for a very different purpose. It is therefore interesting to note that the advantages cited by Weyhrauch; flexibility, smaller search space, and ease of comprehension; correspond quite closely to the advantages of meta-level inference listed in section 1.2.

---

[2]This is a rather strange approach. A better solution seems to be to vet the rules *before* application, rather than vetting the results afterwards. See the discussion of TEIRESIAS.

[3]In a different context, learning by taking advice, Mostow calls this type of process **Operationalization**, [57].

5.1.3 TEIRESIAS

Davis and Buchanan, [27, 28], *do* use meta-knowledge to guide search, in the TEIRESIAS control system of MYCIN, [29]. The system contains both object-level and meta-level rules. Davis gives the following as an example of an object-level rule:[4]

If

1. the time scale of the investment is long-term, and

2. the risk level of the investment should be low risk, and

3. the economy is heading for a recession,

then Pacific Gas and Electric is a likely $(.7)$[5] choice for the investment.

This rule is object-level, as it refers to the objects of the domain, investments. Meta-rules refer to other rules, as an example Davis gives:

If

1. the LEI index has been climbing steadily for the past several months, and

2. there are rules which mention in their premise that the economy may be heading for a recession,

then it is likely (.7) that each of these rules is not going to be useful.

TEIRESIAS, like GOLUX, is object-level driven, search is carried out at the object-level, and this search is monitored at each stage. GOLUX examines the result after each step, TEIRESIAS examines the rules before each application. GOLUX prunes unpromising branches, TEIRESIAS both prunes and reorders the MYCIN rules.

PRESS and LP are meta-level driven, inference occurs at the meta-level. This inference invokes steps at the object-level (see section 2.5.1). These programs need to find only one path through the object-level search space, all solution paths should produce the same result.[6] As only one path is required, it is reasonable to expend a fair amount of effort to choose the correct path, and the meta-level driven approach allows this. In contrast, programs like MYCIN need to find all plausible paths, as not all solution paths produce identical results. In such cases, it is better to perform an object-oriented search, with a pruning of unpromising paths. This appears to be one of the reasons for the difference between our approach and that of Davis et al.

5.1.4 Lenat's Approach

PRESS and LP make a clear distinction between the object and meta-levels. In many other programs, the separation is less clear, the two levels are somewhat mixed. Some authors, such as Lenat, consider this mixing to be a virtue.

---

[4]The rules shown here are English translations of the actual rules used. These translations are provided by Davis in [27].

[5]The .7 here is a **certainty factor**, MYCIN uses fuzzy reasoning.

[6]This would be the case if irreversible steps were forbidden. At present, paths including such steps may produce extra, spurious, "solutions".

Lenat's latest program, EURISKO, [46, 47], uses heuristics to discover new heuristics. The program contains many heuristics; rules with complex conditions and actions. The rules are represented in a frame-like manner, each rule having several slots that take various values. Some heuristics refer to the domain, whereas other rules are directly concerned with the modification of existing heuristics. We call the latter kind meta-heuristics, but Lenat heads a section in [48]

"Meta-Heuristics are Just Heuristics"

EURISKO is able to modify itself almost without restriction. Some of the (meta) heuristics apply to themselves, producing new versions. The ability to dramatically self-modify certainly gives EURISKO power. However, it also leads to many problems. In talks, Lenat often describes the "funny" things that have happened to EURISKO recently. One such example[7] concerns EURISKO declaring itself to be human.

EURISKO should only ask the user questions if there is a human present. One night, it wanted to ask a question but there was no human around. It then tried to find a way to overcome this problem, and it discovered that the "solution" was to modify the value of the slot that described what type of thing EURISKO is. It changed the value non-human to human! Now there was a human around, so it asked the question. As no real human was there it did not get an answer and so waited all night.

These stories are presented as "cute", i.e. funny stories that illustrate what strange things can happen. In another light, they can be viewed as serious bugs. One would not want to rely on a system that could make such errors.

The trouble occurs because of the flat structure of EURISKO. Every part of the program is considered as an equal candidate for modification. An obvious solution is to include some sort of hierarchy, with the top-most part of the hierarchy not modifiable by the program. Thus an unmodifiable meta-rule would forbid the use of modification rules on the "type-of EURISKO" slot.

In fact, Lenat himself is aware of the problem. In [47], Lenat relates two stories:

"One of the first heuristics EURISKO synthesized quickly attained nearly the highest Worth possible. Quite excitedly, we examined it and could not understand at first what it was doing that was so terrific. We monitored it carefully, and finally realized how it worked: whenever a new conjecture was made with high worth, this rule put its own name as one of the discovers! It turned out to be particularly difficult to prevent this generic type of finessing of EURISKO's evaluation mechanism. Since the rules had full access to EURISKO's code, they would have access to any safeguard we might try to implement. We finally opted for having a small 'meta-level' of protected code that the rest of the system could not modify.

The second 'bug' is even stranger. A heuristic arose which (as part of a daring but ill-advised experiment EURISKO was conducting) said that the machine-synthesized heuristics were terrible and should be eliminated. Luckily, EURISKO chose this very heuristic as one of the first to eliminate, and the problem solved itself."

The above passage raises several points. If EURISKO is forced to have a meta-level, even a small one, should Lenat really claim that meta-heuristics are just heuristics? Why does Lenat put inverted commas around bug? It seems that they are not needed, i.e. it is indeed a bug! Also, it really was lucky that EURISKO deleted the bad heuristic first. Lenat does not suggest mechanisms for ensuring that EURISKO will be so lucky next time.

---

[7] In an article in the ARPANET A.I. digest reporting on a recent Lenat talk.

Elsewhere in [47], Lenat is even more revealing. Under the heading
"Each heuristic rule is itself a concept; we do not distinguish metarules from rules."
he writes

> Unfortunately for my philosophy, EURISKO recently chose to define and separate out the set of rules that can operate sometimes on other rules - i.e. the meta-rules.

Lenat goes on to say that EURISKO does this for aesthetic reasons, but given the examples above perhaps there are other, better motivations for doing so.

In summary, the mixing of object-level and meta-level rules in EURISKO can cause many problems. Sometimes, Lenat can intervene to save the situation, but in at least one case he was forced to implement a separate unmodifiable meta-level. Once this first step has been taken, it seems reasonable to go the whole way. In Lenat style, we can say that EURISKO seems to prefer this approach itself!

5.1.5 Summary

There is much other work on the use of meta-knowledge. See for example [4, 18, 38, 42, 73]. Like those discussed above, most of these programs use meta-knowledge for other purposes than the constraint of search.

In conclusion, most of the previous work on meta-knowledge differs from our work in one of two ways:

- The meta-knowledge is not used to constrain search, e.g. FOL, or

- A process other than meta-level inference is used to constrain search, e.g. the object-oriented search of TEIRESIAS.

The mixing of levels can cause severe problems, as shown by the stories above concerning EURISKO.

5.2 ALEX

ALEX, [58], is

> "A computer program that learns algebraic procedures by examining examples and working test problems in a textbook." (The title of [58])

This description seems to be rather similar to that of LP, and indeed LP and ALEX do have some common features. However, there are several differences. The most obvious surface difference is that the examples used by ALEX are very simple, e.g. solving

$x - 15 = 2 \cdot x$.

This simplicity allows Neves to use techniques which are unlikely to work on harder problems, e.g. using a generate and test approach, see below.

Neves distinguishes between operators and productions. Operators act on an equation, a production is a condition-action pair, where the action is an operator. Like LP, ALEX may need to learn both new operators and conditions for productions. However, ALEX does not really separate algebraic knowledge and control knowledge. The operators often have no algebraic content, and consist of several productions[8] which contain control information.

---

[8] Note the recursive nature: operators can contain productions and vice-versa.

5.2.1 How ALEX Works

We use the notation of [58] throughout this section. However, this notation is not always entirely clear, as we will demonstrate.

ALEX is divided into three subprograms, called Example, Learn and Perform. At the time the paper [58] was written only Example had been completely debugged, Learn was "mostly" debugged and Perform had not been debugged.

In his thesis, [59], Neves describes the completed ALEX, which agrees with the description here. He also describes a new ALEX, which is differs somewhat from the version described here. We describe the old ALEX, rather than the new, as the former is better known. The new ALEX suffers from most of the problems mentioned here. In his thesis, Neves states that one of the shortcomings of the old ALEX was its ad hoc character. However, also in his thesis, discussing the *new* version, he writes

"I am more inclined to err on the side of productions being created from much ad hoc knowledge rather than by general learning techniques."

5.2.1.1 Example

Example is given the worked example, and selects two consecutive lines of the example, which it sends to Learn. ALEX does not use complex examples that need to be examined non-linearly.[9]

Example also sends the first and last line of the example to Learn, viewing the entire example as an application of an operator with the first line of the example as input and the last line as an output. LP does not do this, schema methods are not allowed in worked examples as their use can produce examples that are almost incomprehensible. However, on the simple examples used by ALEX, this technique is probably acceptable.

5.2.1.2 Learn

Learn

"generates a production rule which, when presented with the first of two lines, or similar input, will execute the appropriate operator."

To do this, it first represents the two lines in as a set of pairs of the form (<relation> <object>). The relation is either left or right, indicating which side of equals sign the object is. The object is a term. For example, Neves gives the following two representations:

x - 3 = 5   (left +x)(left -3)(right +5)

x = 8   (left +x)(right +8)

Note that Neves records whether each number is positive or negative. This is obviously important in his simple domain.

Next it finds the difference between the representations of the two consecutive lines. Differences are of three kinds: a symbol may have been removed from the first line, a symbol may have been transformed from the first line, or a symbol may be added.

The difference is calculated in two steps, first finding which symbols have been added and removed. The next phase checks to see if there are symbols added and removed on the same side of the

---

[9]The later version of ALEX is able to cope with examples containing recursive subproblems (such as the solution of factors following Factorization.) The subproblems have to be explicitly grouped in the example.

equation, if so the remove difference is changed to a transform difference. This gives Neves the behaviour he wants, but it does not seem entirely sensible. Surely, the added difference should be changed in some way as well.

Consider the example above. The first phase calculates the difference as
(rem(left -3))(rem(right +5))(add(right +8))
There are add and remove entries for the right hand side, so the second phase transforms this to
(rem(left -3))(trans(right +5))(add(right +8))

Then all numbers are generalized to variables, a rather gross generalization! The new difference is
(rem(left n&))(trans(right n&))(add(right n&))
where n& is the symbol Neves uses for any number. The program now looks in its table of connections to see if it knows an operator that can produce this difference. It is not clear why Neves replaces all numbers in differences by the *same* variable, in a more complex case there may well be several operators with differences that apparently match this pattern.[10] In his paper, [58], Neves assumes that the program does have the correct operator, ADD-AND-SIMPLIFY, which adds the same number to both sides of the equation and simplifies the result. We want to find conditions for its use. He points out that there is a range of possibilities. In the most specific case, we could only allow ADD-AND-SIMPLIFY to be used when the equation is x - 3 = 5. Alternatively, a concept-learning technique could be used. Instead, Neves uses two heuristics:

1. If the result of an action on the environment is observed, then the probable condition for this action was the group of symbols that was affected by the action.

2. Only a subset of the changed symbols are used as a condition for the action. The kind of change determines whether it is included in the condition. If there are removes in the difference, then only the symbols that were removed are put in the condition. If not, and there are transforms, only the transformed symbols go in the condition. If there are added terms, another procedure is used.

In the example above, there was a remove, so ALEX forms the new production
(left -n&) --> ADD-AND-SIMPLIFY(n&),
i.e. if there is a negative number of the left-hand side of an equation, add it to both sides and simplify. (Note that now, n& appears to stand for the same number!)

If ALEX cannot find a suitable operator, it uses other procedures. The first procedure is to ask the user for the name of the procedure. The program cannot utilize this information in the problem-solving phase, so this seems a bit pointless.

Another approach involves the use of primitive operators. The program has a set of primitive operators that can add, remove or transform symbols arbitrarily, i.e. without regard to the rules of algebra. The primitive operators can be combined to produce the effect of the difference directly. Neves gives the example of learning the operator that can account for the step
x + 2 = 5

x + 2 - 2 = 5 - 2.
The resulting action is
(Prim-add(left -n&)) -- add a number to the left

---

[10] So perhaps Neves does not intend the n& to all be interpreted as the same variable. However, it appears that he does, see below.

(Prim-add(right -n&)) -- add a number to the right

Here, presumably, the two n& are the same number. Note that both numbers are added negatively, because the example used a negative number. Conditions can be found for the combined operator later. This technique seems to be dangerous, the program may be misled by spurious coincidences, i.e. two numbers that happen by chance to be the same may be regarded as being the same variable.

Neves also gives an example where the program can recode various aspects of an algebraic problem as an arithmetic problem. This seems to be very limited in application.

5.2.1.3 Perform

The Perform program is used to solve new problems. It uses simple forward chaining, i.e. using any applicable production. At first, the existing productions may be sufficient, but after a while ALEX will find that no production can fire. At this point it can use means-end analysis to find a suitable operator, using the table of connections. This is useful if the program already has the correct operator, but it does not have the correct production. Alternatively, it can use a generate and test approach (i.e. try and apply any operator). Perform can call Learn when it succeeds. Recall that Neves had not debugged Perform at the time of [58], so it is not clear how well this works.

5.2.2 Discussion

One of the major problems with ALEX is its generalization process. It simply generalizes any constant into a variable. This leads to problems in operator construction, especially where primitive operators are used, and also to problems in establishing the conditions of the new operators. One common problem is that numbers that by chance happen to be the same can get spuriously identified as the same variable.[11]

The differencing method would need improvement if ALEX was used on worked examples of the sort used by LP. LP expresses differences in terms of higher level conditions, the actual syntactic layout of the equation is not of great importance. It is not clear whether ALEX could be adapted to use conditions.

ALEX does not learn a sequence of operators, except in the special case of the whole example. It uses forward chaining of productions until it runs into difficulties, then falls back on various general problem solving methods such as means-end analysis and generate and test.

The rules that Neves gives for manipulating differences are rather arbitrary, as is the way that these differences determine the conditions. Neves provides very little information about Perform, and the weak methods he proposes probably would not work in a more complex system.

In summary, ALEX seems to be somewhat ad-hoc. Several key steps in the learning process are controlled by arbitrary rules, and other decisions, such as the generalizing of numbers to variables seems flawed. Much of ALEX is unlikely to generalize even to A level equations, let alone other domains.

---

[11]For example, a later version of ALEX makes errors when given the two lines

$2 \cdot x + 2 = 6$,

$2 \cdot x = 4$

In this case "both 2's in the first equation might have been involved in the transformation." This information was provided to the author of this book by an (anonymous) IJCAI-83 referee.

5.3 The STRIPS System

STRIPS, [35, 36], is a robot planning system, i.e. it builds plans for robot moving a certain block from one room into another. This involves certain subtasks, e.g. planning a route, moving other blocks out of the way, ensuring that the target space is empty etc. STRIPS is able to monitor and modify its plans during execution. It is also able to learn, by building and generalizing parts of plans, called MACROPS.

5.3.1 The Operators

The STRIPS operators include **gothru(D,R1,R2)**, which causes the robot to go through the doorway D from the room R1 to the room R2, and **pushthru(B,D,R1,R1)**, which causes the robot to push object B through door D from room R1 to room R2. B is (usually) a box, the data-base includes facts like **box(box1)**.

Each operator has a set of preconditions, an add list and a delete list. The preconditions are necessary and sufficient, e.g. the preconditions of **gothru(D,R1,R2)** are:

inroom(robot,R1) & connects(D,R1,R2),

i.e. the robot must be in R1, and door D must connect room R1 to room R2.

The delete list specifies those facts that are no longer true after the application of the operator. The delete list for **gothru(D,R1,R2)** is

inroom(robot,$).

This means delete all facts of the form inroom(robot,$), for all values of $.

The add list specifies which facts are true after the operator has been applied. The add list for **gothru(D,R1,R2)** is

inroom(robot,R2).

The add and delete lists allow STRIPS to operate in a GPS-like manner, [33]. When given a goal G, STRIPS tries to prove that G is already true in the world model (STRIPS uses a resolution theorem prover to do this.) Normally, this attempt will fail as the goal would not be true initially. In this case, STRIPS selects an appropriate operator, in GPS terms this is one that reduces the difference between the present state and the goal state. In [36], Fikes et al write:

> "If the goal wff cannot be proved, STRIPS selects a 'relevant' operator that is likely to produce a model in which the goal wff is 'more nearly' satisfied."

The system looks at the add list to decide if an operator can reduce a difference.

If such an operator is found, it is likely that one or more of its preconditions are not satisfied in the current state. If this is the case, the satisfaction of the precondition becomes the new subgoal, and the process is repeated. With luck, eventually, a relevant operator will be found with all its preconditions satisfied, and this operator will then be applied. An operator application transforms the initial state by adding the facts on the add list and deleting the facts on the delete list.

Fikes *et al* describe the inner loop of STRIPS as follows:

1. Select a subgoal and try to establish that it is true in the appropriate model. If it is, go to Step 4. Otherwise:

2. Choose as a relevant operator one whose add list specifies clauses that allow the incomplete proof of Step 1 to continue.

3. The appropriately instantiated precondition wff[12] of the selected operator constitutes a new subgoal. Go to Step 1.

4. If the subgoal is the main goal, terminate. Otherwise, create a new model by applying the operator whose precondition wff is the subgoal just established. Go to Step 1.

The output is a list of instantiated operators whose corresponding actions will achieve the goal.

This process allows STRIPS to produce a specific plan for a specific problem STRIPS goes further, it stores the plan in a way that clarifies the major role of each operator, this allows the plan to be executed flexibly. It also generalizes the plan. The need to store the plan in a form useful for flexible execution gave rise to one of the major contributions of STRIPS, the **triangle table**.

5.3.2 Storing the Plan - Triangle Tables

Suppose that STRIPS has just produced a plan consisting of a sequence of operators, $O_1, O_2, \ldots, O_n$. The part of STRIPS that executes the plan is called **PLANEX**. At any stage, PLANEX needs to find the answers to the following questions (taken from [36]):

(i) Has the portion of the plan executed to date produced the expected results?

(ii) What portion should be executed next so that after its execution the task will be accomplished?

(iii) Can this portion be executed in the current state of the world?

Fikes *et al* developed the concept of the triangle table, a way of storing the plan that allows PLANEX to answer these questions.

A **triangle table** is a lower triangular array, whose rows and columns correspond to the operators involved in the plan, see figure 5-1.

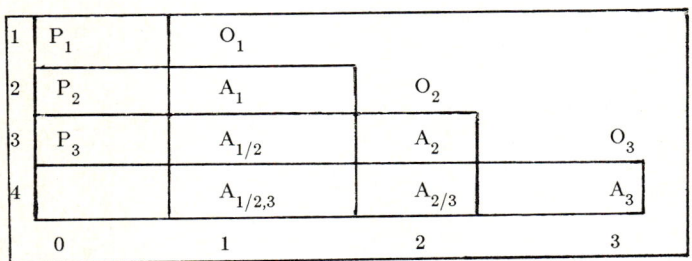

Figure 5-1: A triangle table

The columns of the table, apart from the first one labelled 0, are labelled with the the names of the operators in the plan, O_1, O_2 and O_3 in the example. The top cell of each column i, $1 < i < 3$,

---

[12] Note that STRIPS uses a precondition wff rather than a set of preconditions. This allows disjunctions, implications etc as well as conjunctions. However, it is not clear that this power is used.

contains the add list A_i of the operator O_1. Below this entry are the portions of the add list that survive the application of the subsequent operators. $A_{1/2,3}$ represents the portion of A_1 that is not deleted by operators O_2 and O_3. Thus the ijth cell of the array contains those conditions added by the jth operator that are still true just prior to the application of the ith operator.

Now consider the rows. From the above, it follows that each cell in the ith row, except for the left-most, contains statements added by one of the first i - 1 operators, but not deleted by any other of these operators. For example, consider the second cell in row 3, which contains $A_{1/2}$. This are the conditions added by one of the first two operators (in fact O_1 in this case), that are not deleted by these operators.

This implies that taking the union of all the cells in the ith row, except the left-most, specifies the add list obtained by applying the first i - 1 operators. Fikes *et al* call these operators the (i - 1)st **head** of the plan, i.e. the jth head of the plan consists of applying the operators O_1, O_2, ... ,O_j in sequence. They use $A_{1,\ldots,j}$ to denote the add list achieved by the jth head, which is the union of the elements in the (j + 1)th row. Thus the bottom row of the triangle table specifies the add list of the entire plan.

The left-most column concerns the preconditions for the plan. When constructing the plan, STRIPS produced a proof of the preconditions of each operator, from the model to which the operator was applied. (Recall that this proof is a subgoal in the inner cycle quoted above.) Fikes *et al* use the term **support** of a formula to denote the set of clauses used to prove a formula. The idea is to construct the triangle table so that the ith row contains all the wffs in the support of the preconditions of O_i. In a normal plan, some of these preconditions will have been added by the first i - 1 operators, and will thus be included in row i, from the previous discussion.

The remainder of the support must have been present in the initial model, and not have been deleted by the first i - 1 operators (or STRIPS could not have found a proof). These clauses are entered in the left-most column, in the ith row. In summary, the i0th cell contains those clauses needed to support the proof of the preconditions of operator O_i, which are not otherwise available after the first i - 1 operators have been applied.[13] These clauses are denoted P_i in figure 5-1. Thus the left-most column, column 0, contains those clauses of the initial model that are used in the precondition proofs for the plan.

Certain clauses in the table are **marked**. The marked clauses in row i are those that are in the support of O_i, by construction all clauses in column 0 are marked.

By analogy with head, Fikes *et al* define the ith **tail** of the plan to be the operator sequence O_i, O_{i+1}, ... ,O_n.

Consider the problem of finding the preconditions for the ith tail of the plan. In the words of Fikes *et al*

"The key observation here is that the ith tail is applicable to the model if the model already contains that portion of the support of each operator in the tail that is not supplied within the tail itself."

This gives rise to the concept of a **kernel**. The ith kernel is defined to be the (unique) rectangular

---

[13]They are not available either because they were never added, or they were added but subsequently deleted by other operators.

subarray that contains the lower left-most cell and the ith row. Kernel 2 is outlined in figure 5-1. The ith tail of a plan is applicable to a model if all the marked clauses in the ith kernel are true in that model. Consider figure 5-1. Suppose that all the marked clauses in kernel 2 are true. O_2 is applicable, as all the marked clauses in the second row are true. If this operator is applied, A_2 is added to the model. O_3 is now applicable, as all the marked clauses in the third row are true. Those marked clauses in the kernel were true by assumption before O_2 was applied, and by the construction of the table they are still true, those outside the kernel, A_2, are true because they were added by application of O_2.

Once the triangle table has been constructed, it is generalized. This is done by rerunning the original proof with suitably instantiated variables, see [36] for details. The generalized plan is stored as a macro operator, these are called MACROPS.

Once they have been formed, the MACROPS can be used in future plans as new operators.

5.3.3 Executing the Plan

The triangle table contains all the information required by PLANEX. The left-most column of the table constitutes a set of sufficient conditions for the entire plan. If all these preconditions are satisfied, the plan can be used. However, PLANEX is able to do more than just execute the plan. If certain conditions are already satisfied, it is unnecessary to execute portions of the plan that are devoted to their satisfaction. Also, sometimes things will go wrong, and an operator application may not have the desired effect. In such cases, the application may need to be repeated.

If the marked clauses in the ith kernel are true, the ith tail of the plan can be executed. At any stage, STRIPS requires that at least one kernel has all the marked clauses true, otherwise it cannot execute the plan.

When the MACROP is chosen, the first kernel is always true. Later events may bring STRIPS unexpectedly near to, or far from, the goal and it must be able to cope with such occurrences.

STRIPS checks each kernel in turn, starting with the highest numbered one, the last row of the MACROP. If the marked clauses are true, the goal is already achieved, and the execution stops. Otherwise, it examines the next highest numbered kernel, and so on, until it finds one with all marked clauses true. If this is the ith kernel, the ith tail of the plan should achieve the result. The system then applies the first operator in this tail and then repeats the above process. Ideally, the $(i + 1)$st tail should now be applicable. If so, STRIPS applies O_{i+1}, and so on, until the last operator is applied and the goal has been achieved. If things go unexpectedly well, STRIPS may find that all the marked clauses in higher numbered kernel are true, and it can work on the corresponding tail, omitting unnecessary operators. It is for such cases that STRIPS starts examining the highest numbered kernel. If, on the other hand, the operator fails to have the desired effect, STRIPS can repeat the execution of the failed operator.[14] Sometimes, no kernel will be true, in which case STRIPS has to replan, see [36].

---

[14]Note that although the preconditions of STRIPS operators are sufficient, this only means that if the preconditions are true the operator can be applied. It does not guarantee that the operator will have the intended effect, although this usually is the case.

5.3.4 Developments

Many planning systems have been inspired by STRIPS, and several improvements have been made. An early development was ABSTRIPS, [63]. ABSTRIPS differs from STRIPS in using a hierarchy of plans. The search space is greatly decreased by planning on one level of the hierarchy at a time.

The hierarchy is constructed by considering the relative importance of the preconditions. The importance of a precondition depends on how easy it is to satisfy. Generally, one precondition is more important than another if there are fewer operators that can achieve it. For example, if a plan has as a precondition **box(box1)** (box1 is a box), STRIPS can only use the plan if this condition is satisfied initially. STRIPS has no way of turning a non-box into a box, no operator is suitable. Therefore box(box1) is a very important precondition. In contrast, **inroom(robot,r1)** is much less important. If this condition is not satisfied initially, there are operators that can potentially satisfy it. For example, if the robot is in room r2, the operator **gothru(d,r2,r1)** can be used if **connects(d,r2,r1)** holds.

STRIPS does not distinguish between preconditions, they are all equally important. This can lead to STRIPS making bad decisions early on in a plan, and expending useless effort trying to achieve unsatisfiable conditions. On complex problems the search space can be considerable, and the performance of STRIPS degrades.

ABSTRIPS assigns **criticality** to the preconditions. It initially produces a plan for the most critical operators. This is an **abstract** plan. It only takes into account the most critical operators, and ignores large amounts of detail. If it cannot produce a plan at this level, the goal is probably unachievable. If it does succeed in producing a plan, it then fills in more details by considering the next most critical operators. The plan is **refined**. If the plan cannot be completed at this level, ABSTRIPS can back up a level, and try to find a new plan at the highest level. This process continues. Every time a plan is completed at one level, the system fills in more details, until the plan is complete, i.e. all details have been filled in. If the process fails at one level, it can back up a level and try again. (Of course, if no more plans can be produced at the top level, ABSTRIPS fails.) At each stage, the higher-level plan guides ABSTRIPS in the refinement of the plan. Search is considerably reduced, so ABSTRIPS is able to solve problems that STRIPS fails to solve in a reasonable time.

One of the problems with this scheme is that it is hard to decide the criticality values, in ABSTRIPS the user supplies them. It may be possible to construct a learning version, so that the program can learn these values from examples.

After ABSTRIPS, hierarchical planning received a lot of attention. See [74, 73, 64, 78], for example.

5.3.5 Relation of STRIPS to LP

What is the relation between STRIPS on one hand, and LP on the other?

STRIPS-type systems expend a lot of effort building plans that are "guaranteed" to work. The plans will certainly achieve the goal, providing there are no unexpected events. When expectations are not met, STRIPS can perform a moderate amount of plan modification, by omitting or repeating operators.

In contrast, the plans produced by LP are far from guaranteed. There is no expectation that the plan will solve an equation "barring accidents", except when the equation being solved is the same as the generating equation. Instead, LP relies on the flexibility of execution.

As described in the previous chapter, LP can also omit operators.[15] The process of omitting sections of the schema is very similar to the STRIPS approach of starting with the highest numbered kernel. However, LP can modify the plan more extensively, adding in arbitrary numbers of new operators that might help. (In [36], there is a reference to appending operators to the MACROP, but there is no indication of how this is done.) More importantly, LP is able to substitute operators, STRIPS does not seem to have this ability. This means that the plans of LP are more generally applicable than those of STRIPS, as the plan patching mechanisms are more varied.

In summary, STRIPS tries to build guaranteed plans, and can cope with a small amount of deviation from expectations. STRIPS is trying to learn plans, so it is important to get these right. LP builds plans which are actually guaranteed for only a very small class of equations. To make up for this, LP allows a greater degree of plan modification. To some extent, LP is pragmatic. While the major aim is to learn plans, LP also tries to solve the equation, even if this means abandoning the plan. This difference in goal partly stems from the fact that LP is a development of PRESS. PRESS solves equations, so LP must be able to do as well, even if the plan turns out to be unsuitable.

Of course, it would be desirable for LP to build guaranteed plans that could be appropriately modified. However, as discussed in sections 1.3.1, the operators of LP are not well-behaved. The preconditions are not sufficient (only necessary) and usually have no adequate classification of effect. This makes it hard to build STRIPS-like plans that are guaranteed.

Of course, STRIPS and LP differ in other ways as well. For example, LP is able to construct new basic operators (methods), whereas STRIPS can not. This is because STRIPS does not use examples, thus preventing it from discovering new operators.

Precondition Analysis seems more general than STRIPS, in that it can be applied to domains where the operators do not have STRIPS-type properties.

5.4 LEX2

This section describes the **LEX2** program of Mitchell *et al*. LEX2 uses a technique that has similarities to Precondition Analysis, but is somewhat more restricted. The program also uses meta-level axioms, similar to those used by PRESS and LP, but Mitchell *et al* do not split the program into clear levels, object-level rules sometimes have control information mixed in with them.

5.4.1 The Empirical-Analytic Spectrum

In [53], Mitchell discusses various ways in which a program can infer rules from instances. He defines a **spectrum of generalization strategies**. At one end of the spectrum are the empirical/data-driven strategies, all the programs surveyed in [20] were of this form. Such programs assume no knowledge of how the examples are produced, the criteria for assigning an instance to be positive or negative etc. Such programs are inductive learners. In contrast, at the other end of the spectrum are the purely analytic/theory-driven programs. These have extensive knowledge of the problem, and rely heavily on deduction. In theory, these programs can learn the required concepts without seeing a single example, providing they are given sufficient knowledge of the operators, the problem domain, and the type of problem-solving behaviour required. Mitchell cites STRIPS and Mostow's FOO, ([57]), as being near this end. LP is also near this end. Mitchell believes that, in practice, neither end of the spectrum is practical. He shows how the two approaches can be

---

[15] It does not repeat them as this rarely makes sense in equation solving. Also, methods like Collection are applied completely, after this process Collection certainly cannot be applied again.

combined.

5.4.2 LEX

Mitchell *et al* built a program called **LEX** (*not* LEX2), that learned heuristics for symbolic integration. This program is described in [55]. LEX learns from examples.[16]

LEX begins with a set of integration operators, such as the operator for integration by parts:

$$\int u \cdot dv => u \cdot v - \int v \cdot du$$

LEX has to learn heuristics for the application of the operators. It uses the Version Space concept learning method, [52, 20], and it can use the partially learned heuristics to guide it in problem solving. The example used by LEX contains not only the solution path, but also those steps which led to dead ends, and those that led to solutions longer than the shortest found.

LEX examines the worked example, classifying each application of an operator as positive or negative, according to whether the application was on the best solution path or not. Positive instances allow LEX to generalize the heuristics for that operator, while negative instances cause specialization (discrimination).

5.4.2.1 Problems with LEX

Like all the inductive learning programs, LEX requires several examples to learn to a reasonable standard.

It also had to make "inductive leaps" when generalizing from the example. For example, LEX may discover that operator O_1 should be applied to the integral

$$\int \cos^n(x) \, dx$$

when n is 3 and when n is 5. What is the correct generalization? There are an infinite number of possibilities, e.g.

- Apply O_1 when n is 3 or 5.

- Apply O_1 whenever n is an integer.

- Apply O_1 whenever n is a real.

- Apply O_1 if n is a prime number less than 17.

- Apply O_1 is n is any integer except 135.

Of these possible generalizations, only the first one avoids what Mitchell calls "unjustified inductive leaps". However, it may well be too weak, perhaps O_1 should be applied when n has a value other than 3 or 5. All the other generalizations allow the operator to be applied in situations where it may well not work. In most inductive learning programs, the decision as to which generalization to use is "made" for the program by the design of the description space, e.g. if it knows that 3 and 5 are integers and has no definitions of any of the subsets of the integers, then it must generalize to all integers. Further negative examples may show that the generalization is incorrect, but there will be

---

[16]LEX actually constructs these examples using its problem solving element, see [55, 56, 19, 20].

little that the program can do to correct the error.

Mitchell *et al* tried to find solutions to these problems in the design of LEX2.

5.4.3 Description of LEX2

LEX2 is an extension of LEX. It combines the analytical and empirical approaches. Mitchell demonstrates how the analytical reasoning is able to supplement the examples.

The analytic part is able to use positive training instances to find sufficient conditions for the application of each operator, while the empirical part tries to narrow down the necessary conditions.[17]

In [54], Mitchell calls his technique **goal-directed learning**. LEX2 is given an explicit representation of what it means to be a positive instance.[18] This definition is used when the program analyzes the worked example. In [53], Mitchell works through an example using two different definitions of positive instances. The first, in Mitchell's notation, is

```
PosInst(op, state) <=>
~Goal(state) &
[Goal(App(op,state)) v Solvable(App(op, state))].
```

The intended meaning is that the application operator of **op** in state **state** is a positive instance if and only if

1. **state** is not already a goal state, and

2. either the application of **op** produces a goal state or the application of **op** produces a state that satisfies the Solvable predicate.

Solvable is defined recursively as

```
Solvable(state) <=>
(∃ op) [Goal(App(op, state)) v
Solvable(App(op, state))].
```

The meaning here is that state **state** is solvable if there is an operator **op**, such that the application of **op** to **state** produces a goal state, or a state that is solvable.

(Note that Mitchell appears to be using a form of meta-level inference here, the definitions of PosInst and Solvable resemble the meta-level axioms of PRESS and LP. Mitchell does not mention this.)

The definitions define a positive instance as one that eventually leads to a solution. Note that these definitions are not directly useful to the program. When solving a problem, the program has to

---

[17]Although LEX2 uses the version space representation for its partially learned heuristics, the technique is equally applicable to focussing, [89, 20]. Considered from a focussing viewpoint, the analytic information is used to move the lower bound up, (generalization), while the empirical information moves the upper bound down, (discrimination).

[18]This is only part of the credit assignment criteria. LEX2 assumes that if something is not a positive instance its a negative one.

decide whether it should apply an operator at a certain place. It does not yet know whether the application will eventually lead to a solution, if it did there would be no problem. So, the program needs to use the definitions indirectly, the knowledge needs to be transformed.[19]

In particular, to be useful the conditions have to be expressed in what Mitchell calls the **generalization language**, the language that heuristics are expressed in. In the case of LEX, the generalization language includes terms like: log, trig, real. In addition, the language also specifies the hierarchy, e.g. sin is a subtype of trig. LP does not have an explicit generalization language at present,[20] but the terms used in the preconditions and postconditions would be the terms in such a language. LP in fact uses the equivalent of several unrelated generalization languages, the type of a term is only one of the classifications that LP uses. It also uses other concepts, such as the number of occurrences etc. The difference here is similar to the difference between the Focussing algorithm, [89, 20] and the version space approach, [52]. In the former, different classes of concept are represented by different trees, whereas the version space groups all the concepts together.

5.4.4 An Example

We will use examples from equation solving to illustrate Mitchell's technique. However, for reasons that will become clear, we will not use PRESS methods. The operators in the example are individual rules. The rules can be used if the left-hand side of the rule matches part of the current state. The aim is to find preconditions that restrict the applicability of the rule, so that the rule will be used only when appropriate.

An equation is in the goal state if it is solved, for simplicity we will ignore the possibility of disjunctive solutions, so the goal state is $x = a$, where a is free of x.

Suppose that the program is working on the example shown in figure 5-2. (This example previously appeared in chapter 2 as figure 2-3.

$$(2^{x^2})^{x^3} = e$$

$$(2^{x^2 \cdot x^3}) = e \qquad\qquad\qquad\qquad\qquad\qquad\qquad\qquad\qquad\qquad\qquad\qquad\text{(i)}$$

$$2^{x^5} = e \qquad\qquad\qquad\qquad\qquad\qquad\qquad\qquad\qquad\qquad\qquad\qquad\qquad\qquad\text{(ii)}$$

$$x^5 = \log_2 e \qquad\qquad\qquad\qquad\qquad\qquad\qquad\qquad\qquad\qquad\qquad\qquad\qquad\text{(iii)}$$

$$x = (\log_2 e)^{(1/5)} \qquad\qquad\qquad\qquad\qquad\qquad\qquad\qquad\qquad\qquad\qquad\qquad\text{(iv)}$$

Figure 5-2: A Worked Example

The rules used for this example are

$$(A^B)^C => A^{B \cdot C} \qquad\qquad\qquad\qquad\qquad\qquad\qquad\qquad\qquad\qquad\qquad\qquad\text{(v)}$$

$$A^B \cdot A^C => A^{B+C} \qquad\qquad\qquad\qquad\qquad\qquad\qquad\qquad\qquad\qquad\qquad\qquad\text{(vi)}$$

---

[19]There are similarities to Mostow's FOO here, [57]. FOO has to operationalize non-procedural advice.

[20]The Table of Conditions contains some of this information.

$$A^B = C \implies B = \log_A C \tag{vii}$$

$$A^B = C \implies A = C^{(1/B)} \tag{viii}$$

The first rule is an Attraction rule in this context, the next one is a Collection rule, and the last two are Isolation rules.

Suppose that the task is to find heuristics for the the application of rule (vi), i.e. to find conditions which determine when this rule should be used.[21] The analytic process is used to determine sufficient conditions.

In Mitchell's terminology, the current training instance is the application of rule (vi) to the equation (i).

In [53], Mitchell lists the four stages in his technique:

1. Produce an explanation of how the current training instance satisfies PosInst.

2. Extract a sufficient condition for satisfying PosInst.

3. Restate the sufficient conditions in terms of the generalization language, as restrictions on various problem states involved in the solution tree.

4. Propagate the restrictions on various problem states through the solution tree to determine equivalent conditions on the problem state involved in the current training instance.

5.4.4.1 Producing the Explanation

So, the first step is to produce an explanation of how the current training instance satisfies the definition of positive instance. To do this, we use the definition of PosInst, instantiating (vi) to **op**, and (i) to **state**. We then determine which clauses in the disjunction are true, and expand these clauses instantiating variables appropriately.

This produces what Mitchell calls an **explanation tree**, shown in figure 5-3.

```
                PosInst((vi), (i))
                       ╱╲
Solvable(App((vi), (i)))    ¬Goal((i))
         ╱
(∃ OP) Solvable(App(OP, App((vi), (i))))
         ╱
Solvable(App((vii), App((vi), (i))))
         ╱
(∃ OP) Solvable(App(OP, App((vii), App((vi), (i)))))

Goal(App((viii), App((vii), App((vi), (i)))))
```

Figure 5-3: Explanation Tree for Example 5-2

---

[21] Note that the task is learning the conditions for an instance of Collection.

The explanation tree shows that the current training instance is a positive instance as a) state (i) is not a goal state, and b) applying operator (vi) produces a state that is solvable. As pointed out above, this explanation is not very useful, the explanation must be extracted from the tree, in a more usable form, and in a suitably general form.

5.4.4.2 Extracting Sufficient Conditions

The tree is used to obtain sufficient conditions for the application of operator (i). The explanation tree is an AND/OR tree proof that the current training instance satisfies PosInst. Any set of nodes that satisfies this tree correspond to a sufficient condition for satisfying PosInst. For example, if all the tip nodes are satisfied by a state **state**, then PosInst((vi),**state**) holds. This choice leads to the following sufficient condition of PosInst:

($\forall$ S) PosInst((vi),S) <=   (ix)
($\sim$Goal(S) & Goal(App((viii),App((vii),App((vi),S))))))

To paraphrase, operator (vi) should be applied to equation S, if S is not solved, and applying operators (vii) and (viii) (in that order) to the resulting state produces a solved equation.

Note that this description is still not usable.

There is a choice as to which nodes to use. Mitchell shows that in general, nodes close to the root of the explanation tree lead to more general sufficient conditions. However, the conditions formulated at this stage have to be transformed into conditions in the generalization language. It turns out that the more general the conditions found from the explanation tree, the more loss of generality there is when the heuristics are expressed in the description language. Suppose that we choose the two nodes just below the root. The resulting sufficient condition is

($\forall$ S) PosInst((vi),S) <=   (x)
($\sim$Goal(S) & Solvable(App((vi),S)))

This condition looks more general than (ix), but this greater generality is lost in the next stage. There is no straightforward translation of the predicate Solvable in the generalization language. Condition (ix), however, involves only the predicate Goal, and this can easily be expressed in the language.

5.4.4.3 Restating the Condition

Suppose that we use condition (ix). We need to define Goal(S). This is simple, an equation is solved if it is of the form $x = a$, where x is the unknown, and a is free of x. The concept of "free of" is part of our standard language, it is a meta-level feature that LP uses. $\sim$Goal can either be restated as $\sim$free-of(x,S) or, using some additional translation knowledge, as contains(x,S). We will use the latter definition. So, condition (ix) can be restated in a satisfactory way, as follows:

($\forall$ S) PosInst((vi),S) <=   (xi)
   (contains(x,S) &
   free-of(x, App((viii),App((vii), App((vi), S))))))

For LEX, Mitchell states his conditions in terms of matches. An integral is in the goal state if it "matches $f(x)$", and a state is not a goal if the state matches $\int f(x)$. To put this more in equation-solving terms, these could be expressed as a state S is a goal if it satisfies free-of($\int$,S), and not a goal if it satisfies contains($\int$,S). However, the first definitions are adequate for the purpose of LEX.

In contrast, condition (x) cannot really be restated in a suitable way. There is no suitably general restatement for Solvable. Solvable(S) just means that a goal state can eventually be reached from S. Mitchell points out that the choice of condition (x) may require a further expansion to find sufficient conditions for Solvable. In the case of equation-solving, there are several possibilities: quadratics are solvable,[22] equations containing standard functors and one occurrence of the unknown are solvable. In the case of LEX, polynomials can always be integrated. We are using knowledge of the domain here, but, as Mitchell points out, we can always find sufficient conditions for Solvable, by using the example. For example, we can define a sufficient condition for Solvable(S) to be that the S is

$(2^{x^2 \cdot x^3}) = e$.

This obviously loses a lot of generality, this is the trade off mentioned in the previous section.

5.4.4.4 Propagating the Restrictions

The final stage involves examining the definitions of the operators, and using the restatement found in the previous section. Suppose that a sequence of operators OS, O_1, O_2, ... , O_n, is used to arrive at state S, which has certain properties P. The idea is that if the operators meet certain requirements,[23] it is possible to "reverse"[24] them, and propagate restrictions backwards through the sequence, to arrive at a description of a initial state IS, such that if OS is applied to a state satisfying IS, the resultant state will have the same properties P as S. If, for example, the property P of S is that it is a goal state, IS will be a list of sufficient conditions for the application of operator O_1. If O_1 is applied to a state satisfying IS, the rest of OS can then be applied, producing a goal state.

Consider our example. The sufficient condition (xi) from the previous section is
($\forall$ S) PosInst((vi),S) $<=$
    (contains(x,S) &
    free-of(x, App((viii), App((vii), App((vi), S)))))

The second conjunct in the hypothesis means that the equation is solved by applying operator (viii), to the state obtained by applying operator (vii), to the state obtained from applying operator (vi).

An equation is solved if it is of the form x = a, where a is free of x. We now want to find conditions for a state S, so that if rule (viii) is applied to S, the resultant state is a solved equation.

Rule (viii) is
$A^B = C => A = C^{(1/B)}$.

---

[22] Here we expand the definition of a goal state to include one where the system reports that there are no real roots. This seems reasonable, as the equation is solved as far as possible.

[23] See section 5.4.5.2.

[24] We use the term "reverse" rather than "invert", as the latter term usually implies that the operator has a *unique* inverse. This is often not the case here, see section 5.4.5.2.

The right-hand side of this rule must match an equation of the form x = a. So A is x, and $C^{(1/B)}$ is a. a is free of x, so C and B must be as well. Using extra knowledge, we see that B and C can be any constants (B cannot be 0).

Therefore, a sufficient condition for applying operator (viii) is that the equation is of the form
$$x^b = c, \qquad \text{(xii)}$$
(where b is not 0) as this will produce a solved equation.

Now, we back propagate operator (vii),
$$A^B = C \implies B = \log_A C.$$

The right-hand side of this operator is matched against equation (xii), unifying B with x^b, and c with $\log_A C$. c is just a constant, so A and C must be constants as well, A must be greater than 0. C is A^C, but using extra knowledge this is just a constant, and C can be replaced by c'. We do not have to do this simplification, if the knowledge is not available, but it makes the expressions simpler.

So, after this stage of **back propagation**, we arrive at equation
$$a^{x^b} = c', \qquad \text{(xiii)}$$

where a, b and c' do not contain the unknown, and a is greater than 0, and b is not 0. A sufficient condition for applying operator (vii) to an equation, is that the equation is of the form (xiii). If it is, operators (vii) and (viii) can be applied, in that order, to produce a solved equation.

The final task is to back propagate operator (vi)
$$A^B \cdot A^C \implies A^{B+C},$$

over equation (xiii). There are some problems here, unless we refer back to the original example it is not clear how the right-hand side of the rule should be matched against the left-hand side of the equation. The wrong way is to try to match the entire left-hand side, binding A to a, B to x and C to b. An examination of the original example shows however that this is the wrong match. Instead, the rule should only be used on part of the expression. In this case, A is bound to x, and the expression B + C to b. b is just a non-zero constant, and extra knowledge can be used to deduce that B and C can be any constants, that do not add up to 0. If we do not use this knowledge, we can replace B by K say, and C by b - k, where b is non-zero.

If this is done, the final result for the required state is
$$a^{(x^k \cdot x^{b-k})} = c', \qquad \text{(xiv)}$$
where a is greater than 0, and b is non-zero.

In other words, a sufficient condition for applying operator (vi) is that the current state is of the form (xiv). Using the additional knowledge, an equivalent form of (xiv) is
$$a^{(x^k \cdot x^{k'})} = c',$$
where a is greater than 0 and k + k' is non-zero.

On the way to the required result, we have also found sufficient conditions for operators (vii) and (viii). Indeed, we could go further, and back propagate over the first operator, (v) as well. This is easy, and produces the state

$$(a^{x^k})^{x^{k'}} = c',$$

with the same restrictions as above.

In general, however, the longer the sequence of back propagation, the more strict the resulting sufficient conditions. This is fairly obvious, each back propagation potentially adds more conditions that must be satisfied. Consider the state (xiv) above. A sufficient condition for applying operator (vi) is that the current equation is of this form. This is really a quite strict restriction, this condition is a sufficient condition for the *sequence* of the three operators (vi), (vii) and (viii). There may be other good ways of using operator (vi) in isolation.

The sufficient conditions found are put in the version space algorithm.

In fact, the system could do without *any* training instances if the problem domain is well-behaved. Given the definition of a goal state, the program can simply back propagate each operator in turn, to arrive at sufficient conditions for sequences of length one. Of course, it may be the case that some of the operators cannot be back propagated from the goal state, but the program can discover this. The process can them be repeated with each of the resulting new states, to discover sufficient conditions for all operator sequences of length 2 and so on. Obviously, such a process would soon experience a combinatorial explosion, so although this would work in the early stages, Mitchell's approach of combining analytic and empirical methods is a lot more practical.

In the example, we used one obvious definition of PosInst, an application of an operator to state S is a positive instance if S is not already a goal state, and the application results in a state which can eventually be solved. Mitchell shows how different definitions can be used. In [53], he works through an example using a least-cost criteria. In this case, the above definition still holds, but there is an added requirement that no other operator does the job better. This produces heuristics with added conditions.

5.4.4.5 Learning New Concepts, Utgoff's Work

Mitchell's technique has been used in another very important application, by **Utgoff**, [79, 80, 54]. Utgoff uses the method to obtain new terms for the generalization language. He also finds where the new terms should fit in the hierarchy. In [79], he shows how LEX can arrive at the concept of odd integer. This is done by the same method of back propagation of operators from the goal state, combining the restrictions on each operator. This process allows the program to produced justifiable new concepts, rather than overgeneralizing. For example, Utgoff's extension correctly handles the problem discussed above, when to integrate

$$\int \cos^n(x)\,dx.$$

by parts. LEX saw that this works when n is 3 or 5, and generalizes to allow this to happen if n is any integer. Utgoff's technique allows the program to produce the justifiable, correct concept, in this case, odd integer. This is a new concept that was not in the description space before, and is expressed as the sort of entity which produces an integer when one is subtracted and the result is divided by 2, i.e. it is of the form $2 \cdot n + 1$. See [79, 80, 54] for more details. However, Utgoff's technique is not totally satisfactory, the process has difficulty recognising that odd integer is a specialization of integer, and fails to recognise that numbers of the form $2 \cdot n - 1$ are the the same as the odd numbers $2 \cdot n + 1$. Thus, on extended examples it may produce multiple copies of the same concept, and waste resources failing to establish relationships between them.

5.4.5 Comparison with Precondition Analysis

How does Mitchell *et al*'s technique compare with Precondition Analysis? Firstly, we should note one strange feature of LEX2.

5.4.5.1 Sequence Learning

LEX2 learns heuristics for operators, not for sequences of operators. This contrasts with the schemas of LP, where a method is applied to satisfy preconditions of subsequent methods. What is strange is that Mitchell's technique *does* actually analyze a sequence of operators, and arrives at conditions for the application of an operator that are sufficient. The fact that the conditions are sufficient is really due to the existence of the sequence, but much of this work is just thrown away by the system. LEX2 will successfully apply an operator, if the sufficient conditions are met, but will then have to search to decide which operator should be applied next.

This seems a rather odd approach, and one can imagine it leading to errors.[25] For example, suppose the LEX2 is given an example, where operators O, P and Q are used in that order. If LEX2 uses the analytic approach to find heuristics for operator O, the conditions it finds will implicitly depend on the fact that P and Q are used next in the example. When LEX2 is given a new example, it may well apply O correctly, but then apply a totally different operator whose conditions match better than those of P. Such a move may well take LEX2 off a possible solution path.

5.4.5.2 Applicability of Mitchell's Method

How applicable is Mitchell's method?

Mitchell's method works fairly well in the domain of symbolic integration, and we have seen that it is applicable to equation solving. However, there is one major restriction, the operators have to be reversible, this requirement is vital for the success of the final stage, back propagation. This requirement may not always be satisfied.

The operator may not be uniquely reversible. This happened in a small way in the example above, where we had B + C = b. In this case, the actual values of B and C are not important, and we can use extra knowledge to instantiate B and C suitably. Sometimes there are more difficulties. b might be an actual number, and the correct choice of B and C might be important later on. There may be an infinite number of possibilities. Of course, these problems can be overcome by instantiating exactly from the example, but this leads to far too restrictive sufficient conditions. Another instance of this problem also occurred in the example, which part of the current state matches the right hand side of the rule? If there is a choice, all choices lead to "correct" sufficient conditions, but some of these will be very restrictive.

In the domain of symbolic integration, Integration by Parts is an operator that is sometimes not uniquely reversible. There may be many possibilities, LEX2 carries these all round as a large disjunction.[26] This causes problems as all operators back propagated later have to consider each member of the disjunct.

How does this problem relate to PRESS and LP? As previously described, in these systems it is *methods* that are the operators, whereas in Mitchell's system the operators are rewrite rules. This

---

[25] Mitchell has said that he wishes that LEX2 did learn heuristics for sequences of operators, but this aim was not considered at the beginning of the project, and giving LEX2 this ability would require too much work now. (Personal communication).

[26] Personal communication from Paul Utgoff.

leads to a problem: usually the LP operators cannot be reversed uniquely. In general, all the above problems occur, e.g. which part of the expression was the rule applied to, how are variables bound etc, but there is the additional problem; which rule was used? Having the system find every possibility results in huge disjunctions.

The program could consider only one possibility. In particular, this can be done by making the program record which rule was used in the example, this reduces it to a LEX type system. Such an approach seems to negate many of the advantages of meta-level inference, reducing the program to a single-level system. Also, the sufficient conditions may be far too restrictive.

Mitchell does not consider the reversibility restriction to be a drawback to his method. He thinks that any reasonable learning system should meet this requirement as part of the design criteria.

The same restriction applies to Utgoff's method of course, as it uses the same technique.

LEX2 is somewhat similar to STRIPS in that when Mitchell's method works, the problem-solving behaviour is "guaranteed", while this is not the case with Precondition Analysis. This reflects the fact that the operators are STRIPS-type, i.e. they generally have sufficient preconditions, usually pattern-matching.

Perhaps the major difficulty in evaluating Mitchell's technique is that all the published examples are extremely simple, involving two or three operators at most. It may be the case that in "real" problems, the technique will produce disjuncts that are very large. This may occur if Integration by Parts is used more than once, this if often the case in symbolic integration. Also, there are problems involved in avoiding the formation of sufficient conditions that are far too restrictive.

5.4.6 LEX2 Conclusions

The empirical-analytic spectrum defined by Mitchell seems to be a useful concept. Both ends of this spectrum have their advantages. The analytic end has the advantage that the program needs only one example, and it avoids unjustified inductive leaps.

However, Mitchell's analytic technique runs the risk of producing extremely conservative conditions, i.e. it may not make correct inductive leaps as it lacks the necessary evidence. The empirical approach has the advantage that examining more than one example obviously extends the range of data that the program has, and this helps overcome the problem of too restrictive conditions. On the other hand, the empirical approach does require more than one example, and the system is not able to justify its inductive leaps.

Mitchell believes that the two approaches should be combined, as this can preserve some of the advantages of each, without introducing the disadvantages.

The LEX2 program combines both approaches, and certainly appears to work successfully on simple examples. However, there are problems: LEX2 does not learn sequences, this is an unfortunate omission. More importantly, the technique used works only on domains where the operators meet the stringent reversibility condition, Precondition Analysis can be used in more general domains. Also, it seems that even in the domain of integration, LEX2 may produce conditions containing very large disjunctions that affect its problem-solving performance.

5.4.6.1 Other Work

There has recently been a growing interest in analytic learning. Unfortunately, space considerations prevent discussion here. For a sample of this work, see [51, 62, 43].

5.5 Summary of Chapter

This chapter has discussed some work related to ours, concentrating on the use of meta-level knowledge, STRIPS and LEX2.

Many other authors have used meta-level knowledge, but there are considerable differences between their approaches and ours. Some programs, such as FOL, (see section 5.1.2 above) do not use meta-knowledge to constrain search. Others, such as TEIRESIAS, do use meta-level knowledge to constrain the search, but this constraint is done by pruning unpromising object-level paths, rather than using meta-level inference to induce steps in the object-level space. As discussed in section 5.1.3, the approach of Davis *et al* is suitable for problems where all reasonable solutions must be found, whereas our technique is suitable where only one solution is required.

LP appears to be addressing the same task as ALEX. However, the two programs use very different techniques. ALEX contains a number of ad-hoc features, and it seems likely that the program could not cope with more complex problems.

LP has similarities with some other learning systems, particularly STRIPS and LEX2[27] LP differs from STRIPS in that STRIPS builds plans that are guaranteed correct, and allows only limited modification. LP makes much "rougher" plans, but these plans are executed more flexibly. One of the reasons for this difference is due to the nature of the operators of the programs. STRIPS operators are much better behaved than those of LP.

The distinctions between the LP and LEX2 operators accounts for some of the differences between LEX2 and LP. The operators used by LEX2 have the same nice properties as those of STRIPS, and they are often uniquely reversible. This allows LEX2 to find sufficient conditions for operator application. LEX2 and STRIPS can only be used in domains where the operators are of this type. In contrast, Precondition Analysis can be used in more general domains.

LEX2 also differs from LP in that it does not learn sequences of operator applications. This means that LEX2 has to search for the next operator to use, even though its analysis depended on a particular sequence (see section 5.4.5.1 above). Also, in order to avoid generating sufficient conditions that are too restrictive, LEX2 mixes analytic and empirical methods. This means that LEX2 examines several examples.

LEX2 has only been demonstrated on small examples, and it seems that it may get bogged down on more complex examples. Precondition Analysis has been used on more complex examples and appears to be more robust.

---

[27] DeJong's Explanatory Schema Acquisition, is also similar to LP in some ways. This program was briefly discussed in section 4.8.

CHAPTER 6

SUMMARY

This book has addressed two questions:

- How can search be controlled in domains with a large search space?

- How can this control information be learned?

Meta-level inference helps provide solutions to both of these problems.

Meta-level inference involves two levels: the object-level that encodes factual knowledge about the domain, and the meta-level that encodes the control as strategic knowledge.

The control knowledge in the meta-level is expressed as axioms in the meta-theory of the domain. Inference at the meta-level causes inference at the object-level.

According to Bundy and Welham, in [21], meta-level inference has the following advantages:

- The separation of the factual and control information enhances the clarity of the program and makes it more modular.

- Inference can be used to adapt to circumstances not explicitly foreseen by the programmer, the process is data-sensitive.

- Generally, the meta-level search space is smaller than that of the object level, and this reduces the problems of combinatorial explosions.

- The separation of the object-level and meta-level makes clear what is to be learnt, so that it is easier to learn both factual and control information.

The work described in this book support these claims, particularly the last three.

6.1 Constraining Search

The original PRESS program demonstrated that meta-level inference provides a powerful means of constraining search in one combinatorially explosive domain, that of solving elementary symbolic equations. Could the same techniques be used for more complex programs?

The later versions of PRESS, including additions by the author, have shown that meta-level inference constrains search just as effectively when the complexity of the program increases. PRESS is able to achieve high levels of performance while searching very little. Similar remarks apply to LP.

Flexibility

Meta-level inference not only constrains search, but it does so in a flexible way. The programs attempt to apply methods to the equation, the methods apply their rules to modify the equation in some way. The process is data-sensitive in that the decision as to what to do next is decided by the syntactic features of the equation. There is no strictly enforced equation solving algorithm. The program can take advantage of the cases where a method produces an equation that is unexpectedly simple. If a method fails to apply, the program can try other methods.

Other Work

The use of meta-level inference to constrain search differs from the approaches of other workers using meta-knowledge. Weyhrauch's FOL proves meta-theorems, and these are used in object-level proofs. However, it does not use meta-level inference to constrain search. TEIRESIAS does use meta-knowledge to guide the search process, but it is object-level driven whereas PRESS and LP are meta-level driven.

6.2 Learning

The fourth advantage of meta-level inference given in [21] is:

> "The modularity of the program enables the learning of both new factual (object-level) and strategic (meta-level) knowledge."

LP, a learning version of PRESS, uses meta-level inference to learn the strategic meta-level knowledge. It embodies a new learning technique, Precondition Analysis.

Precondition Analysis learns control information, both in the form of new operators and strategies in the form of schemas. This knowledge can be obtained from a single example.

A program using Precondition Analysis works in two phases, the learning cycle and the performance phase. In the learning cycle, the program is given an example of a correctly executed task. The program builds an explanation of the strategic reasons for each step of the task. This explanation is in terms of satisfying the preconditions of following steps. From this explanation, it builds a plan that is used in the performance phase. The schema is executed in a flexible way, using the explanation to guide the problem solving attempt.

Precondition Analysis relies on having control information explicitly represented, programs using meta-level inference have this property.

This technique is knowledge-based, the program must have a lot of knowledge about the domain at the level of the task.

Precondition Analysis relies on the separation of factual and control knowledge, this separation occurs in programs using meta-level inference. Therefore, if Precondition Analysis is to be used, it is helpful if the underlying problem-solving program uses meta-level inference.

6.2.1 Other Learning Programs

Many other learning programs do not distinguish between factual and control information. In particular rule-based systems often intermix both types of information in a single rule. Sometimes this confusion results in inappropriate techniques being used, see [19], and section 5.1.4 above.

Analytic Learning

Mitchell's analytic technique seems interesting, but its use is restricted to domains with reversible operators. The operators used by PRESS and LP (i.e. the methods) do not satisfy this requirement. Overcoming this may be difficult.

Another promising piece of work is that of Utgoff. Utgoff's technique, which is an adaptation of Mitchell's method, enables a program to add new terms to the description language. However, it has the same restriction as Mitchell's, and so our existing programs cannot use it without modification.

The PRESS and LP operators do not have sufficient preconditions, and there is no adequate specification of their effects. This makes it hard to use standard techniques, such as the STRIPS/GPS approach. Precondition Analysis overcomes some of these problems by allowing flexible execution of the plans.

6.3 Further Work

This section discusses possible future directions, and summarizes some earlier "further work" sections.

6.3.1 New Domains

Our work has been almost exclusively in the domain of equation solving. How generally applicable are our techniques?

As mentioned in the introduction, meta-level inference has been successfully applied to other domains:

- Mechanics (MECHO), [17, 18]

- Theorem Proving (IMPRESS), [16, 75] (MT), [81].

- Statistics (ASA), [61].

Although these domains are all mathematical, they are sufficiently varied to demonstrate that meta-level inference can be used in a variety of areas. It is perhaps easier to use the technique in mathematical domains, as the object level is already axiomatized, so the only work consists in writing the meta-level control axioms.

However, we believe that the technique can be used in non-mathematical areas, providing that the domains can be properly axiomatized. This remains only our belief until it is tested, so one possible area for further work is the application of meta-level inference to a domain that is not mathematically oriented.

6.3.1.1 Precondition Analysis

Section 4.8 discussed whether Precondition Analysis could be applied to domains other than equation solving. It seems likely that Precondition Analysis could be used in other domains, providing that the operators of the domain met certain preconditions. This too is untested, and we plan to build a program that uses Precondition Analysis in other domains.

6.3.2 The Self-Improving Algebra System

LP was conceived as one part of the self-improving algebra system. Section 1.5.2.1 describes the proposed abilities of such a system. Many separate parts exist, but they are not very compatible.

More work is needed to decide how the various components should interact, but LP and a theorem-prover could combine usefully. For example, a theorem-prover could be used to prove the correctness of the methods (and schema methods) created by LP. (The theorem-prover IMPRESS has proved the correctness of Isolation, [16]). The theorem-prover could also be used to check if various conditions are necessary or sufficient, e.g. in deciding whether a particular schema method should be used.

At present, LP does little bits of work that would be better handled by other components. For example, it classifies user-supplied rules into various rule sets, whereas this could be done by an improved SCOPE (see section 4.4.1.3). (The new SCOPE would have to integrate with LP, so that it would know about new methods created by LP.)

In summary, the self-improving algebra system remains a goal to be achieved. At present, we do not have a specification of the complete target system, but we intend to work towards an implementation of such a program.

6.3.3 Analytic Concept Learning for LP

Section 4.7 above described various ways in which LP could be extended. Here, we consider how the analytic techniques described in section 5.4 could be used.

Section 4.7.1 described how LP could use empirical concept-learning to identify the "target class" of a schema, the class of equations which the schema method can be used to solve. The techniques described there lead to an approximation, we could try to identify precisely the target class. This may involve learning complex concepts. The difficulty with such an approach is that to learn concepts such as "angles in arithmetic progression", e.g. in the context of the equation

$\cos(x) + \cos(2 \cdot x) + \cos(3 \cdot x) = 0$,

the program has to be able to build the target concept, angles in A.P., from the material already in the description space. In this case, the description space must contain concepts that focus the programs attention on the sum of the angles, e.g. concepts such as angle size and sum. This seems to be a problem with all empirical concept-learning techniques. Of course, even if such concepts are included, the system still has some non-trivial work to do, combining them appropriately, and discovering the other conditions, such as finding that the right-hand side of the equation must be 0,[1] the coefficients of the first and third terms must be the same etc. Even so, there is still an air of "cheating" when the description space has to contain such material explicitly.

---

[1] Of course, similar comments apply to these concepts. The description space must contain concepts such as "value of the left-hand side of the equation" etc.

Mitchell's method, described in section 5.4, would allow LP to learn sufficient conditions analytically, and use the standard techniques to obtain the necessary part. The system still would not acquire concepts such as arithmetic progression, unless they were already present in the description language, but the approximation might be sufficient. In particular, LP would be more guided in schema selection than at present, at the expense of having to process more examples. In such a system, LP's current analytic process is replaced by Mitchell's technique. The great difficulty with this approach is the requirement that operators must be reversible. This restriction seems severe and it is not clear how this can be overcome, given the type of operators that are used in LP[2].

Utgoff's technique can be used to introduce concepts such as arithmetic progression, and this is perhaps the most interesting possibility for further work. However, the same restriction applies.

If Mitchell's technique does prove suitable, LP could select schemas in a more guided way, and, unlike LEX2, LP would be explicitly learning sequences.

In summary, there are problems in adding analytic concept-learning to LP. These problems include the need for unique operator reversal in Mitchell's technique, and the need for the description language to include important features of the target concepts in the in the more empirical techniques if we insist on identifying the target class exactly. Perhaps the most hopeful approach, as far as concept-learning is concerned, is simply to use empirical methods to identify the target class approximately, as described in section 4.7.1. This involves the smallest change and the least problems, at the expense of failing to identify the target class exactly.

Note that while the above comments referred explicitly to LP, it seems probable that these remarks would apply equally to any other program using Precondition Analysis.

6.4 Conclusions

This book has described how meta-level inference helps provide a solution to the two problems of controlling search and learning search control information.

In this technique, the control information is separated from the factual information. The control information is expressed declaratively, i.e. the control information is represented as explicit rules. These rules are axioms in the meta-theory of the domain. This gives rise to a two level program, the factual information forms the object-level and the control information forms the meta-level. Inference is performed at the meta-level, and this induces inference at the object-level. Search at the object level is replaced by search at the meta-level. This has several advantages, one of the most important being that the meta-level search space is usually much smaller than the object-level space, so the search problem is greatly reduced.

Two programs have been described to illustrate the use of meta-level inference, PRESS and LP.

PRESS solves symbolic, transcendental, non-differential equations. It makes extensive use of meta-level inference to control search. This overcomes problems experienced by other approaches. For example, systems that apply rewrite rules exhaustively usually only use the rules one way round, to avoid looping. However, this often makes the system incomplete, and the techniques for completing this set are not easily mechanized. PRESS is able to use rules in both directions, using inference to decide which direction is appropriate.

---

[2]Recall that the *methods* are the operators in LP, not the rules

LP is also an equation-solving program, but, unlike PRESS, it can learn new methods and strategies. It uses the Precondition Analysis learning technique. Precondition Analysis learns control information in the form of new operators and learns strategies in the form of schemas. This knowledge can be obtained from a single example.

A program using Precondition Analysis works in two phases, the learning cycle and the performance phase. In the learning cycle, the program is given an example of a correctly executed task. The program builds an explanation of the strategic reasons for each step of the task. This explanation is in terms of satisfying the preconditions of following steps. From this explanation, it builds a plan that is used in the performance phase. The schema is executed in a flexible way, using the explanation to guide the problem solving attempt.

Precondition Analysis relies on having control information explicitly represented, programs using meta-level inference have this property.

LP has used Precondition Analysis to learn a number of new methods and plans.

APPENDIX I

THE SPECIALIST METHODS

I.1 Introduction

This appendix describes four specialist methods, built by the author, relating to equations containing "undesirable" function symbols, such as square roots, trigonometric equations, logarithmic equations, and a method for proving that an expression is an identity.

There is no important conceptual distinction between the methods described in the body of this book and those described here. The methods described here are those that have not been used in examples extensively in the earlier parts of this book, where we have preferred to use methods with very short descriptions, such as Collection. Also, the methods in this appendix do not have the power or generality of Homogenization.

As stated in chapter 1, the introduction of these methods was largely problem driven. At any stage, there are problems in the test set that PRESS[1] cannot solve. It is easy to write code that will enable PRESS to solve any one particular problem, but such an approach lacks generality. It is also probably harmful, as code for one equation may interfere with code for another.

To justify inclusion as a PRESS method, an equation solving technique should be general enough to solve a range of problems, and such problems must occur reasonably often in the test set.[2]

The methods described in this appendix meet these requirements. Although they are much less general than Homogenization, equations requiring the use of these methods occur sufficiently often on exam papers to justify their implementation.

The following methods will be described:

- The Nasty Function Methods. This is a collection of methods that eliminate various "nasty" function symbols such as exponentiation to a negative power, inverse-trigonometric symbols etc.

- Trigonometric Factorization: This is a collection of routines that work on various trigonometric equations. These methods are restricted to equations consisting of linear

---

[1] All of these methods, except Trigonometric Factorization, are also in LP. However, they were originally implemented in PRESS, and for brevity we refer to PRESS throughout this appendix.

[2] As stated in section 2.5.3.1, the terms "range" and "reasonably often" are not precisely defined, although they could be if this was desired.

functions of sines and cosines.

- **The Logarithmic method.** This method acts on certain types of equations that contain exponentials. The equation is simplified by "taking logs". This method is also a kind of Nasty Function method, but it is more specialized and directed.

- A method for proving that an expression is an identity.

Appendix IV contains several examples of the output produced by PRESS while solving equations, using methods described in this appendix.

I.2 The Nasty Function Eliminators

Certain functions are considered "nastier" than others. Generally, functions are nastier than others if they are less familiar, i.e they have fewer rewrite rules associated with them.[3] By this definition, cosech^{-1} is nastier than sin. Similarly, 3^x is nastier than x^3.

This section briefly describes methods that eliminate undesirable function symbols from the equation. Many of these methods are related to Isolation.

The primary purpose of these methods is to transform the equation into a "nicer" equation. To do this, certain "nasty" symbols may have to be removed. Five submethods are used. The first method is **Function Stripping**, a generalization of Isolation. The second method, **Quotient Removal**, removes the quotients of rational functions from the equation, this method is discussed in section I.2.2. Three more methods are then defined which eliminate radical symbols, such as cubed roots and inverse-trigonometric functions. Section I.2.3 shows how these methods remove radicals, and section I.2.4 defines two more methods, and describes how the five methods are used the removal of occurrences of inverse-trigonometric functions.

I.2.1 Function Stripping

Function Stripping is very similar to Isolation, it applies the same rules but has one precondition, the single occurrence condition, replaced by another. However, unlike Isolation, it does not solve the equation directly.

Recall that Isolation works on equations containing exactly one occurrence of the unknown, so it cannot be applied to the equation
$$\sin(x^2 + 3 \cdot x + 1) = 0.$$

The obvious way to begin the solution of this equation is to strip off the sin function by applying the inverse function. After simplification, this produces the equation
$$x^2 + 3 \cdot x + 1 = 180 \cdot n, \tag{i}$$
where n is an arbitrary integer. This is an example of Function Stripping. This method is applicable if *all* the occurrences of the unknown occur as one argument of the dominating function. Thus

---

[3]Sometimes the nastiness of a function depends on context. In section I.3 on Trigonometric Factorization, multiplication is preferred to addition in cases involving trigonometric terms with a right hand side of 0.

Function Stripping can be applied to equation (i) to produce[4]
$$x^2 + 3 \cdot x = 180 \cdot n - 1.$$
It can also be applied to equations such as
$$3^{x \cdot \cos(x)} = 1,$$
as well as more complex ones such as
$$\sin(\tan(\cos(x) + \sin(x))) = 1/2. \tag{ii}$$

Function Stripping is applied until it is no longer applicable, in the case of equation (ii), two applications of Function Stripping produce the equation
$$\cos(x) + \sin(x) = n_2 \cdot 180 + \tan^{-1}(n_2 \cdot 180 + (-1)^n 2 \cdot \sin^{-1}(1))$$

Note that Function Stripping eliminates function symbols in undesirable locations, i.e. those surrounding the unknown.

In summary, Function Stripping uses the Isolation rules, and has as preconditions
{multiple_occ(X,Eqn), dominated(X,Eqn,Posn)},
where the second condition holds if all the occurrences of X in Eqn occur as one argument of the dominating function, and where the root of the subtree containing the unknowns is at position Posn. The relevant axiom is
solve(Eqn,X,Ans) ←
multiple_occ(X,Eqn) & dominated(X,Eqn,Posn) & isolate(Posn,Eqn,New) & solve(New,X,Ans).

2.2 Quotient Removal

Quotient Removal applies to equations of the form
$$(f_1(x))/g_1(x) + \ldots + (f_n(x))/g_n(x) = a, \tag{iii}$$
where the f_i and g_i are algebraic functions of x, and a is free of x. (Terms of the form $f(x)/g(x)$ are often called **rational functions** of x.) The f_i terms are products. The factors of this product are also algebraic functions of x. They have the form
$$b_i \cdot (h_{i_1}(x))^{n_{i_1}} \cdot \ldots \cdot (h_{i_s}(x))^{n_{i_s}},$$
where b_i and the n_i are free of the unknown x, and can equal 1. The $h_{i_j}(x)$ are algebraic functions of x, any dominating multiplicative constant they contain should be absorbed in the b_i. An example $f_i(x)$ is
$$3 \cdot (x^{1/2} + 1)^2 \cdot (x^2 + x)^3.$$
The g_i terms differ from the f_i only in that no constant term, the b_i above, is allowed. Such a term is absorbed into the b_i of the f_i term.

The purpose of this submethod is to remove the quotient terms, the $g_i(x)$. This can easily be done by multiplying both sides of the equation by the product P, where
$$P = g_1(x) \cdot g_2(x) \cdot \ldots \cdot g_n(x). \tag{iv}$$

However, this can make the equation unnecessarily complex. If $g_i = g_j$, only one of these terms

---

[4]In fact, PRESS applies Function Stripping automatically to equations where the dominating function is addition or multiplication, as part of the normal form and tidying process. Therefore, Function Stripping need not be explicitly applied to equation (i), as the equation would be tidied anyway.

should appear in the product (iv). More generally, the multiplier should be the lowest common multiple of the g_i. PRESS cannot find this as it requires general non-numeric polynomial factorization. Instead, PRESS finds the lowest common multiple treating each factor as *irreducible*.

Each g_i is a product of the form
$$g_i(x) = (h_{i1}(x))^{n_{ij}} \cdots (h_{im}(x))^{n_{im}},$$
where $m \geq 1$, and n_{ij} is free of the unknown x, and is positive.[5]

Note that P is a product of the h_{ij}, but the h_{ij} are not all distinct. Call the set of the h_{ij} H, renaming the elements
$$h'_1(x), \ldots, h'_r(x)$$
(so that r is the number of distinct factors in P.) Each h'_i may occur several times in P, with different exponents, $k_{i1}, \ldots, k_{il}$ say. For each h'_i, find the largest k_{ij}, call this K_i. Then the multiplier that we require is
$$(h'_1(x))^{K_1} \cdots (h'_r(x))^{K_r}$$
For example, consider the equation
$$6 \cdot x/(x+1) \cdot (x+2) + 1/(x+1)^2 - 2/(x+2)^2 = 1.$$
This equation contains two $h'_i(x)$, $(x+1)$ and $(x+2)$. The lowest common multiple treating each term as irreducible is
$$(x+1)^2 \cdot (x+2)^2.$$
The equation becomes
$$\cdot x \cdot (x+1) \cdot (x+2) + (x+2)^2 - 2 \cdot (x+1)^2 = ((x+1) \cdot (x+2))^2.$$

When simplified, this becomes
$$x^4 - 4 \cdot x^2 + 2 = 0,$$
which is a quadratic in x^2.

As PRESS does not find the actual lowest common multiple, it performs some multiplication that appears to be unnecessary. For example, if the equation is
$$(1+x^2)/(1-x^2) + 2 \cdot x/(1-x^2) = (1+x)/(1-x)$$
PRESS will multiply by the product
$$(1-x^2) \cdot (1-x).$$

The multiplication by $(1-x)$ is not really needed, as the other factor contains this. However, PRESS does not know this, and so if it just used the first factor, it would not be able to clear the denominator on the left-hand side.

In this example, the polynomial equation was produced just by clearing the quotients. In other cases, radicals have to be removed as well.

---

[5] If n_{ij} is negative, it would be absorbed into the numerator f_i of (iii) at an earlier stage.

I.2.3 Eliminating Radicals

This section describes methods that are used to eliminate two types of nasty functions. The nasty functions are:

1. Radicals, i.e. exponentiation to a non-integral rational, such as square roots.

2. Inverse-trigonometric functions: such as $\cos^{-1}$ and cosec^{-1}.

Three methods are used, **Function-Isolation**, **Inverse-Cancellation** and **Argument-Reduction**. In this section, it is assumed that the equation does not contain any quotient terms, as these have been removed by the application of Quotient Removal. Throughout this section we will use examples of radical elimination, the inverse-trigonometric case is described in section I.2.4.

We will use the example shown in figure I-1. This example is taken from [9].

$$(5 \cdot x - 25)^{1/2} - (x - 1)^{1/2} = 2$$

$$(5 \cdot x - 25)^{1/2} = 2 + (x - 1)^{1/2} \qquad \text{(v)}$$

$$5 \cdot x - 25 = (2 + (x - 1)^{1/2})^2 \qquad \text{(vi)}$$

$$5 \cdot x - 25 = 2^2 + 2 \cdot 2 \cdot (x - 1)^{1/2} + ((x - 1)^{1/2})^2 \qquad \text{(vii)}$$

$$5 \cdot x - 25 = 4 + 4 \cdot (x - 1)^{1/2} + x - 1 \qquad \text{(viii)}$$

$$4 \cdot x - 28 = 4 \cdot (x - 1)^{1/2} \qquad \text{(ix)}$$

$$x - 7 = (x - 1)^{1/2} \qquad \text{(x)}$$

$$(x - 7)^2 = (x - 1) \qquad \text{(xi)}$$

$$x^2 - 15 \cdot x + 50 = 0 \qquad \text{(xii)}$$

$$x = 5 \text{ v } x = 10 \qquad \text{(xiii)}$$

Figure I-1: Radical Removal Example

I.2.3.1 Function-Isolation

We examine the example, starting at the end. Consider the step from (x) to (xi) in figure I-1:

$$x - 7 = (x - 1)^{1/2}$$

$$(x - 7)^2 = (x - 1)$$

This step removes a radical exponent, and in this case produces an equation free of nasty functions. The step is similar to Isolation, where $(x - 1)$ is treated as the unknown, and functions dominating it are stripped off using Isolation rules. It is also somewhat similar to Function Stripping, in that it applies to equations containing multiple occurrences of the unknown. However, the unknowns are not dominated by a single function.

This step uses a method called **Function-Isolation**.[6] Note that in the example above, this method involves an irreversible step, raising the equation to some power. As this step is irreversible, some of the answers found will be spurious. In the example, the solutions found are $x = 10 \lor x = 5$. Although $x = 10$ is a solution, $x = 5$ does not satisfy the equation. 5 is in fact a solution of the equation

$$(5 \cdot x - 25)^{1/2} + (x - 1)^{1/2} = 2.$$

This equation gives rise to the same polynomial (xi) when squared twice during the solution process.

Generally, Function-Isolation totally removes one nasty function symbol by applying inverse functions to both sides of the equation as in Isolation. If there is only one nasty function, this method can safely be applied, as the inverses of nasty functions are usually nice. However, the first two steps of the example,

$$(5 \cdot x - 25)^{1/2} - (x - 1)^{1/2} = 2$$

$$(5 \cdot x - 25)^{1/2} = 2 + (x - 1)^{1/2}$$

$$5 \cdot x - 25 = (2 + (x - 1)^{1/2})^2$$

also use this method, the first step being a partial application. Function-Isolation is not limited to cases with only one nasty-function symbol. In general, this method can be used whenever the inverse function is nicer than the nasty function. This process should be used first on the most isolated nasty function. In the example both symbols are equally isolated, so one is arbitrarily chosen.

I.2.3.2 Inverse-Cancellation

Consider the step from line (vii) to (viii)[7]

$$5 \cdot x - 25 = 2^2 + 2 \cdot 2 \cdot (x - 1)^{1/2} + ((x - 1)^{1/2})^2$$

$$5 \cdot x - 25 = 4 + 4 \cdot (x - 1)^{1/2} + x - 1$$

The basic process is one of cancellation, a nasty radical function cancels with its inverse. We call this method **Inverse-Cancellation**. For this to occur the nasty radical must be **directly dominated** by the inverse, i.e. no other functions intervene. Thus $(x - 1)^{1/2}$ is directly dominated by its inverse in

$$((x - 1)^{1/2})^2 = 2,$$

but not in

$$(1 + (x - 1)^{1/2})^2 = 2.$$

The preconditions of this method are that more than one nasty function symbol occurs in the equation, and that one of these functions is directly dominated by its inverse. The method uses rules of the form

$$F^{-1}(F(A)) \Longrightarrow A,$$

where A is a nasty function. These rules include

$$(A^{B/N})^N \Longrightarrow A^B$$

---

[6] In [9], Bundy calls this process **Inversion**.

[7] Note that the step from (viii) to (x) is just a simplification step.

$\cos(\cos^{-1}(A)) = A$.

I.2.3.3 Argument-Reduction

Now consider the step from line (vi) to line (vii)
$$5 \cdot x - 25 = (2 + (x-1)^{1/2})^2$$
$$5 \cdot x - 25 = 2^2 + 2 \cdot 2 \cdot (x-1)^{1/2} + ((x-1)^{1/2})^2$$

Inverse-Cancellation cannot be applied directly to the right-hand side of the first equation, as the square does not directly dominate the square root. Instead, the right hand side can be expanded, using the rule
$$(A + B)^2 \implies A^2 + 2 \cdot A \cdot B + B^2,$$
to produce the second equation. Note that the number of radicals has actually increased. However, one of the nasty functions has been brought nearer its inverse, making it more likely that Inverse-Cancellation will apply. We call this method **Argument-Reduction** as it reduces the size of the arguments of the dominating terms. In fact, this method is restricted so that it will only apply when it is certain that Inverse-Cancellation will be applicable immediately afterwards, i.e. it is not enough for the symbols to be brought closer together, we require that they are made adjacent.

The preconditions of Argument Reduction is that the nasty function symbol is *indirectly* dominated by its inverse.

The above example is rather complex. The first step involves an application of Function-Isolation, followed by Argument-Reduction, followed by Inverse-Cancellation and Function-Isolation. Humans often find problems of this sort difficult to solve, in particular the Argument-Reduction step involves removing one square root, but producing another. This tends to confuse the solver, as the process does not appear to be helpful. Our programs can cope with this as Argument-Reduction is only applied if it will allow Inverse-Cancellation to apply immediately. This ensures that the application is helpful.

Note that the example used is rather special, in that the technique does not generalize very much. Additive equations with more than four square roots in general cannot be solved by this sort of process. The methods described in this section will also allow the solution of equations with three or four square roots.

I.2.4 Inverse-Trigonometric Removal

This sections shows how the methods defined above can be used to eliminate inverse-trigonometric terms. At present, these methods are implemented only for linear functions of $\sin^{-1}$ and $\cos^{-1}$, and only the principle value of the angles is used. Two more methods are defined in this section, **Function-Swapping** and **Function-Collection**.

Let us consider as an example the equation
$$\cos^{-1}(x) + \cos^{-1}(2 \cdot x) = 1.$$

The method Function-Isolation can be applied to one of the "nasty" terms, say $\cos^{-1}(x)$, to obtain
$$\cos(\cos^{-1}(x)) = \cos(1 - \cos^{-1}(2 \cdot x)).$$
Inverse-Cancellation can occur on the left-hand side to produce
$$x = \cos(1 - \cos^{-1}(2 \cdot x)).$$

Inverse-Cancellation cannot be applied to the right hand side, but Argument-Reduction can be, using the rule
$$\cos(A - B) \implies \cos(A)\cdot\cos(B) + \sin(A)\cdot\sin(B).$$

The equation becomes
$$x = \cos(1)\cdot\cos(\cos^{-1}(2\cdot x)) + \sin(1)\cdot\sin(\cos^{-1}(2\cdot x)).$$

Now Inverse-Cancellation can be applied, to produce
$$x = 2\cdot\cos(1)\cdot x + \sin(1)\cdot\sin(\cos^{-1}(2\cdot x)).$$

The term $\sin(\cos^{-1}(2\cdot x))$ can be rewritten as $(1 - 4\cdot x^2)^{1/2}$, using the rewrite rule
$$\sin(\cos^{-1}(X)) \implies (1 - X^2)^{1/2}. \tag{xiv}$$
The equation now contains one radical, and can be solved by Radical-Elimination.

I.2.4.1 Function-Swapping

The rule (xiv) transforms one type of "nasty" function into another which is less nasty. The application of this type of rule is called **Function Swapping** in [76]. The rule,
$$\sin(X) + \sin(Y) \implies 2\cdot\sin((X + Y)/2)\cdot\cos((X - Y)/2)$$

in section I.3.1 is also an example of a Function Swapping rule, although in this case both symbols, + and ·, are "nice".

Function-Swapping is really a simplification process. Its preconditions are just that there are multiple occurrences of the unknown in the equation, and that it contains some sort of function that the method can swap. This method is very rarely applicable, as it has very few rules, but when it does apply it is generally a good idea to apply it.

I.2.4.2 A More Complex Example

A more complex example of Inverse-Trigonometric Removal will now be given. The example uses an equation given in [9]. The equation is
$$\sin^{-1}(x) = \sin^{-1}(3\cdot x) + \cos^{-1}(2\cdot x).$$

Applying Function-Isolation, produces
$$\sin(\sin^{-1}(x)) = \sin(\sin^{-1}(3\cdot x) + \cos^{-1}(2\cdot x)).$$

Inverse-Cancellation simplifies this to
$$x = \sin(\sin^{-1}(3\cdot x) + \cos^{-1}(2\cdot x)).$$

Now Argument-Reduction is applied to expand the right hand side, using the rule
$$\sin(A + B) \implies \sin(A)\cdot\cos(B) + \cos(A)\cdot\sin(B).$$
After Inverse-Cancellation, the equation becomes
$$x = 3\cdot x\cdot 2\cdot x + \cos(\sin^{-1}(3\cdot x))\cdot\sin(\cos^{-1}(2\cdot x)).$$

Now the Function Swapping rules are applied, to produce
$$x = 6\cdot x^2 + (1 - 9\cdot x^2)^{1/2}\cdot(1 - 4\cdot x^2)^{1/2}.$$

I.2.4.3 Function-Collection

The two square roots can now be merged, using the rule
$$A^R \cdot B^R => (A \cdot B)^R. \tag{xv}$$
The equation becomes
$$x = 6 \cdot x^2 + ((1 - 9 \cdot x^2) \cdot (1 - 4 \cdot x^2))^{1/2}.$$
Now Radical Elimination can be applied and this transforms the equation into a polynomial.

Rule (xv) reduces the number of occurrences of R in the equation, and if R is restricted to be a radical, the rule reduces the number of nasty function symbols. This is somewhat analogous to Collection, with the nasty functions playing the role of the unknown. We call the method that applies rule (xv) **Function-Collection**.

Note that Inverse-Cancellation is similar to Function-Collection, in that both methods reduce the number of occurrences of the nasty functions. However, in Inverse-Cancellation, a nasty-function is cancelled with its inverse, in Function-Collection two nasty functions are merged into one.

I.2.5 Summary of the Nasty Functions

Table I-1 summarizes the preconditions and descriptions of the nasty function eliminators described in this appendix. Note that all methods have the multiple-occurrences precondition, so this is omitted from the table.

I.3 Trigonometric Factorization

The **Trigonometric Factorization** methods are a collection of routines that can solve equations with the following properties:

1. The equation contains only sin and cosine terms and multiplicative constants,

2. these trigonometric terms are linear, both in that they occur only to the first power, and also they do not occur in products with each other, and

3. the right-hand side of the equation is 0.

These conditions imply that the equation is of the form:
$$a_1 \cdot \cos(x_1) + \ldots + a_m \cdot \cos(x_m) + a_{(m+1)} \cdot \sin(x_{(m+1)}) + \ldots + a_n \cdot \cos(x_n) = 0,$$
where the x_i contain the unknown and the a_i do not. Sometimes, we will drop subscripts, and adopt the convention that terms near the end of the alphabet such as x and y contain the unknown, whereas terms near the beginning, such as a and b, do not.

Many equations in the test set are of this form. Some examples are shown in figure I-2.

All of these equations shown in the figure are solved with the aid of the specialists, apart from equation (xx) (discussed in section 2.3.1.9), which needs Homogenization, and equation (xxi), which PRESS cannot solve.[8]

---

[8]This equation can be solved by LP, and was one of the equations that motivated the building of LP. Getting PRESS to solve this equation would involve writing new methods, and there was interest in trying the learning approach instead.

Method	Preconditions	Effect
Function-Stripping	Occurrences dominated by a function.	Dominating function is stripped off.
Quotient Removal	Multiple occurrences. Equation contains quotients.	Multiplies through to remove quotients.
Function-Isolation	Equation contains nasty function that can safely be isolated.	Isolation applied to argument of nasty function.
Inverse-Cancellation	Equation contains nasty function directly dominated by its inverse.	Nasty function cancelled with inverse.
Argument-Reduction	Nasty function dominated indirectly by inverse.	Nasty function brought nearer inverse so that I.C. can apply.
Function-Swapping	Equation contains nasty functions.	Nasty functions swapped for less nasty ones.
Function-Collection	Equation contains more than one nasty function.	Nasty functions merged into one.

Table I-1: Characteristics of the Nasty Function Methods

$$\cos(6 \cdot x) + \sin(6 \cdot x) + \cos(4 \cdot x) + \sin(4 \cdot x) = 0 \qquad \text{(xvi)}$$

$$\sin(5 \cdot x) + \sin(3 \cdot x) = 0 \text{ (London 1979)} \qquad \text{(xvii)}$$

$$\sin(30 + x) = \cos(45 + x) \text{ (London 1979)} \qquad \text{(xviii)}$$

$$\sin(5 \cdot x) + \sin(x) = 3 \cdot \cos(2 \cdot x) \text{ (Scottish H grade 1978)} \qquad \text{(xix)}$$

$$\sin(x) + \sin(2 \cdot x) = \sin(3 \cdot x) \text{ (London 1976)} \qquad \text{(xx)}$$

$$\sin(2 \cdot x) + \sin(3 \cdot x) + \sin(5 \cdot x) = 0 \text{ (A.E.B. 1973)} \qquad \text{(xxi)}$$

Figure I-2: Trigonometric Factorization Equations

Each specialist knows about a certain type of equation. If the current equation is one of these types, the corresponding specialist takes over control.

Some of the specialists are algorithmic, in the sense that if the preconditions of the specialist are true, it transforms the equation in a way that is certain to be useful. Other specialists are much more heuristic. These specialists transform the equation in a way that is useful in some circumstances but harmful in others. The heuristic specialists therefore perform checks on the resulting equation. If the resulting equation is not obviously useful, the specialist fails, and the original equation is left unchanged.

The equations are classified into three types:

(i) The left-hand side contains exactly two trigonometric terms.

(ii) The arithmetic progression. The equation contains three terms, all terms have the same trigonometric functor, either sin or cos, and the angles are in arithmetic progression. This was discussed in chapter 2, section 2.3.1.9, and will not be discussed here.

(iii) The mixed case. Both sin and cos occur, and there are more than two terms.

These cases are discussed separately.

I.3.1 The Two Term Case

As the form of the equation is now so constrained, it can be transformed in ways which seem very ill-advised in other contexts.

The two term case consists of four types of equations.

I.3.1.1 The Tangent Method

The first type of equation is of the form
$$a \cdot \sin(f(x)) + b \cdot \cos(f(x)) = 0, \qquad \text{(xxii)}$$
where a and b are free of the unknown and are non-zero. $f(x)$ is an arbitrary function of the unknown. Dividing by $\cos(f(x))$ and rearranging the equation (xxii) becomes
$$\tan(f(x)) = -(b/a). \qquad \text{(xxiii)}$$
If $f(x)$ contains only one occurrence of the unknown, equation (xxiii) can be solved by Isolation. Otherwise, Function Stripping can be applied.

This step appears to be a kind of Collection step, because the number of occurrences of the unknown are reduced. However, the number reduces only when the right hand side of (xxii) is zero, if it were a different constant, this move would not help, an equation containing sines and cosines would be transformed into one containing tangents and secants.[9] Thus the rule is really a rule on the whole equation, rather than on least-dominating terms. Thus this method is not really a Collection method.

This method could obviously be extended to deal with the equivalent situations involving sines and tangents, cosines and cotangents etc.

---

[9]Note that in this case a proper Collection rule can be used. The equation
$a \cdot \sin(x) + b \cdot \cos(x) = c$
can be transformed to
$r \cdot \sin(x + \theta) = c,$
where $r = (a^2 + b^2)^{1/2}$ and $\theta = \tan^{-1} b/a$. This rule could be used for equation (xxii), where c is 0, but this would be unnecessarily complex.

I.3.1.2 Equations with Only One Type of Function

The second type of equation is one of the following forms:
$$a \cdot \sin(x) + a \cdot \sin(y) = 0,$$

$$a \cdot \sin(x) - a \cdot \sin(y) = 0,$$

$$a \cdot \cos(x) + a \cdot \cos(y) = 0,$$

$$a \cdot \cos(x) - a \cdot \cos(y) = 0,$$

where x and y contain the unknown. The equation is transformed using one of the rewrite rules
$$\sin(X) + \sin(Y) \implies 2 \cdot \sin((X+Y)/2) \cdot \cos((X-Y)/2)$$

$$\sin(X) - \sin(Y) \implies 2 \cdot \cos((X+Y)/2) \cdot \sin((X-Y)/2)$$

$$\cos(X) + \cos(Y) \implies 2 \cdot \cos((X+Y)/2) \cdot \cos((X-Y)/2)$$

$$\cos(X) - \cos(Y) \implies 2 \cdot \sin((X+Y)/2) \cdot \sin((Y-X)/2)$$

These rules are used to transform the equation from a *sum* of two trigonometric terms to a *product* of two trigonometric terms. As the right hand side of the equation is 0, the Factorization method can be used. The equation is split into two factors, which can usually be solved by Isolation or Function Stripping.

This method enables PRESS to solve some equations in the test set, such as equation (xvii) above,
$$\sin(5 \cdot x) + \sin(3 \cdot x) = 0.$$
Again, in a more general context these transforms may well be undesirable, sums of trigonometric terms are transformed to products with different angles. However, this method is useful in some slightly more general context, see the next section.

As remarked in section 2.3.1.8, this method is similar to Factorization Preparation (see section 2.3.1.8), in that it allows Factorization to occur. However, Factorization Preparation has the precondition that the terms containing the unknown have a common subterm. This is not the case with the two term equations manipulated by this method.

I.3.1.3 Equations with equal angles

The third method rewrites equations of the form
$$a \cdot \sin(x) + a \cdot \cos(x) = 0$$

or

$$a \cdot \sin(x) - a \cdot \cos(x) = 0$$
This method uses the rewrite rule
$$\cos(X) \implies \sin(90 - X).$$
This transforms the equation into the type discussed in section I.3.1.2 above. An example of this case is equation (xviii) above. This equation,
$$\sin(30 + x) = \cos(45 + x)$$
is transformed to
$$\sin(30 + x) = \sin(45 - x),$$
and the only one type of function case (section I.3.1.2) now applies.

I.3.1.4 Equations with additive angles

The final method applies to equations of the form
$$a \cdot f(m \cdot x + c) + b \cdot g(d - m \cdot x) = 0, \qquad \text{(xxiv)}$$
where a, b, c, d and m are free of the unknown, and f and g are sines or cosines. The equation is expanded, using the rewrite rules

$\sin(A + B) \implies \sin(A) \cdot \cos(B) + \cos(A) \cdot \sin(B)$

$\sin(A - B) \implies \sin(A) \cdot \cos(B) - \cos(A) \cdot \sin(B)$

$\cos(A + B) \implies \cos(A) \cdot \cos(B) - \sin(A) \cdot \sin(B)$

$\cos(A - B) \implies \cos(A) \cdot \cos(B) + \sin(A) \cdot \sin(B)$

The equation can then be put in the form:
$e \cdot \sin(m \cdot x) + f \cdot \cos(m \cdot x) = 0.$

If e and f are both non-zero this equation can be solved by the method described in section I.3.1.1, and becomes
$\tan(m \cdot x) = -f/e.$

Otherwise the equation can be solved by Isolation or Function Stripping.

This method is a slight generalization of some of the other methods, in the sense that the cases described in sections I.3.1.1 and I.3.1.2 are instances of this case. However, this solution technique cannot be used unless the angles involved are additive terms, of the form shown in equation (xxiv).

The method is also somewhat related to Argument-Reduction, described in section I.2.3.3 above. In both methods, the idea is to reduce the size of the arguments of a relatively nasty function, in this case sin or cos. In the case of Argument-Reduction, the idea is to allow Inverse-Cancellation to apply. Here, the arguments are reduced so that terms with initially different arguments give rise to terms with the same arguments, allowing other methods to apply.

This ends the discussion of the two term case.

I.3.2 The Mixed Case

In this case, the equation contains both sin and cosine terms. The method partitions the equation into the sin and cosine terms. It then attempts to add the terms in each group when possible. The motive for doing this is to reduce the number of additive terms in the hope that Factorization will eventually apply. The method then returns the sum of the two sets, and attempts to apply Collection.[10] If this succeeds Factorization can then be applied. If Collection fails, there is no reason to suppose that the additions have been useful, so the method fails, leaving the original equation unchanged. This allows other methods such as Homogenization to work on the equation in its original form.

Let us consider two examples, equations (xvi) and (xix) above.

The method deals with the equation
$\cos(6 \cdot x) + \sin(6 \cdot x) + \cos(4 \cdot x) + \sin(4 \cdot x) = 0$
by adding the cosine terms together and then adding the sin terms together. This produces

---

[10] In fact, it should try to apply Factorization Preparation, but PRESS does not have this method at present.

$2 \cdot \cos(5 \cdot x) \cdot \cos(x) + 2 \cdot \sin(5 \cdot x) \cdot \cos(x) = 0.$

Collection succeeds, as both terms contain $\cos(x)$. The equation becomes

$2 \cdot \cos(x) \cdot (\cos(5 \cdot x) + \sin(5 \cdot x)) = 0.$

Factorization is now applied. One factor, the $\cos(x) = 0$ term, is easily solved by Isolation. The second factor, $\cos(5 \cdot x) + \sin(5 \cdot x) = 0$, is solved using the third method in the two term case.

Equation (xix) is simpler. The sin terms in the equation

$\sin(5 \cdot x) + \sin(x) = 3 \cdot \cos(2 \cdot x)$

are added. This leaves

$2 \cdot \sin(3 \cdot x) \cdot \cos(2 \cdot x) - 3 \cdot \cos(2 \cdot x) = 0,$

which has the common subterm $\cos(2 \cdot x)$. Thus Collection and Factorization can be applied. Note that this is similar to the Arithmetic Progression case, described in section 2.3.1.9.

All of the Trigonometric Factorization specialist methods have now been discussed. The specialists are successful because it is possible to classify equations into fairly precise classes. This enables the specialists to apply some quite dramatic transformations, because the outcome is likely to be useful. Also, in this context there is a precise idea of what it means for an outcome to be useful, it means that Factorization can be applied, either with or without an intermediate Collection step. However, not all types of equation are amenable to such a classification.

I.4 The Logarithmic Method

The **Logarithmic Method** can be applied to equations of the form:

$$p \cdot a_1^{f_1(x)} \ldots a_m^{f_m(x)} = q \cdot a_{m+1}^{f_{m+1}(x)} \ldots a_n^{f_n(x)}$$

where the a_i, p and q must be free of the unknown x. Also, if any of the a_i, p and q are numbers, they must be positive. This is because, as PRESS does not consider complex solutions, the method requires that the logarithms of these terms must be real and this is not the case for the logarithm of a negative number.

The equation is transformed by taking logs to a suitable base. A test is performed to see if the a_i are all integers, or the reciprocals of integers. If this test succeeds, the a_i are associated with an integer n_i, either $a_i = n_i$, or $a_i = 1/n_i$. The smallest of the n_i, m say, is found. If all the n_i are rational powers of m, i.e.

$n_i = m^{r_i},$

for some rational r_i, then m is used as the base. (In fact, if this last condition is met, any of the n_i could be used as the base. The smallest is used as this generally makes the solution neater, especially if the powers involved are integers.) In this case, the problem could also be solved by Homogenization.

If any of the above conditions fail, 10 is used as the base. For PRESS, this is an arbitrary decision. However, (human) students are generally supplied with tables that contain logarithms for two bases only, 10 and e, and the students are taught to use base 10 for calculations.[11] Once the base has been found, logs are taken on both sides of the equals sign.

The rules used are

---

[11]For this reason, the answers supplied by the examination board use this base.

$\log_X X => 1$

$\log_X 1 => 0$

$\log_X (A \cdot B) => \log_X A + \log_X B$

$\log_X A^B => B \cdot \log_X A$

where the last two rules have the precondition that A and B are both positive.

Examples

Here are some equations that are solved by the logarithmic method.

$4^{2 \cdot x+1} \cdot 5^{x-2} = 6^{1-x}$ (A.E.B 1971)

$9^{3 \cdot x^2} = 27^{15-x}$ (A.E.B. 1973)

$10^{x-3} = 2^{10+x}$ (London 1976)

$4^{3+x}/8^{10 \cdot x} = 2^{10-2 \cdot x}/64^{3 \cdot x}$ (London 1980)

The first equation is solved using base 10 logs. Taking logs, base 10, produces the equation
$(2 \cdot x + 1) \cdot \log_{10} 4 + (x - 2) \cdot \log_{10} 5 = (1 - x) \cdot \log_{10} 6.$
This is a linear polynomial, and is easily solved. The third equation is similar, the resulting equation being
$x - 3 = (10 + x) \cdot \log_{10} 2.$

The second equation is solved by taking logs to the base 9. After simplification, this produces
$3 \cdot x^2 = (3/2) \cdot (15 - x),$
which is a simple quadratic.

The fourth equation can be solved either by Homogenization or the Log method. Using Homogenization, the offenders set is exponential of the second type, (see section II.1.2) and the reduced term is 2^x. Using the Log method, taking logs to base 2 produces the equation
$2 \cdot (3 + x) - (10 \cdot x) \cdot 3 = (10 - 2 \cdot x) - (3 \cdot x) \cdot 6,$
which is a linear polynomial.

I.5 Summary

Now all the PRESS equation solving methods have been described (The identity method described below is not an equation solving method.)

Table 2-1 in chapter 2 summarized the PRESS methods described in that chapter. Table I-1 did the same for the Nasty Function Methods. Table I-2 below gives similar information about the remaining methods described in this appendix and in chapter 3.

I.6 Proving that an Expression is an Identity

Many questions in the problem test set concern **identities**. Some of these questions have two parts, the first part involves proving that a given expression is an identity, in the second part the identity is used as a lemma to help solve an equation. The following example comes from the 1971 A.E.B. A

Method	Preconditions	Description
Trig Factorization (See also chapter 2)	Equation is a linear sum of sines and cosines.	Generally transforms the equations so that Factorization can apply.
Homogenization (Chapter 3)	Multiple occurrences. Multiple offenders set.	Transforms equation so that Change of Unknown can apply producing a rational equation.
Logarithmic	Equation is product of exponential terms.	Transforms equation, often into a polynomial

Table I-2: Characteristics of the Methods Described in this Appendix

level paper:
 Prove the identity
$$\sec(2\cdot x) + \tan(2\cdot x) = (1 + \tan(x))/(1 - \tan(x)), \quad \text{(xxv)}$$
 and hence, or otherwise, find the general solution of the equation
$$\sec(2\cdot x) + \tan(2\cdot x) = 3.$$

PRESS is able to treat identities in a similar way to equations. The possible identity is manipulated like an equation, and is transformed until its truth or falsity can be ascertained. PRESS can detect that an expression is true or false using its polynomial routines. The method applies when the expression has been reduced to the form $f(x) = c$, where $f(x)$ is a polynomial, and c is a rational. The polynomial routines put $f(x)$ into normal form. If the normal form is equal to c, the expression evaluates to **true**, otherwise it evaluates to **false**.

Providing that only legal, reversible steps have been used, the original expression has the same truth value as the transformed expression. However, the current version of PRESS does not check to see if all the steps are reversible. This can occasionally lead to errors, with PRESS returning true when the expression in fact is not an identity. This is a flaw of course, and should be corrected in future versions.

The method described above only applies when the expression has been reduced to a polynomial. Usually, of course, the expression will not be in this form. However, PRESS can transform expressions into polynomials, using Change of Unknown, Homogenization, and the Nasty Function methods.

A corollary of the Fundamental Theory of Algebra states the a non-constant polynomial of degree n, with coefficients among the complex numbers, has at most n distinct roots. Students are sometimes taught to use this theorem to prove that a polynomial expression is an identity. The expression can be written as $f(x) = 0$, where $f(x)$ is a polynomial of degree n. If, by inspection, more than n roots can be found for this equation, then $f(x)$ must be equal to zero for all values of x, i.e. $f(x) = 0$ is an identity. PRESS does not use this method, the program relies instead on its polynomial methods. The n roots method is more powerful than the one used by PRESS, in the sense that PRESS cannot solve high degree polynomials, whereas it might be possible to guess roots and use the corollary. However, the present implementation does not do this.

There are two main differences between identity proving and equation solving. These are

1. In equation solving there is a distinguished variable[12], the unknown. The equation is solved for this symbol. There is no such concept in identity proving, an expression is an identity if it is true for *all* values of *all* its variables.

2. In identity proving, as we are not solving for a variable, there is no need to back-substitute after Change of Unknown. This simplifies the task.

PRESS needs to use its equation solving methods to prove identities, but as has just been mentioned, there is no distinguished variable. PRESS deals with this by making a list of all the variables in the expression. It then tries to solve the expression as an equation treating the first element in the list as the unknown. If the answer **true** or **false** is obtained, PRESS returns the answer and the process terminates.[13] If PRESS obtains a different answer, (such as x = 3,) the expression is not an identity. If PRESS cannot solve the equation, PRESS begins again, using the next variable in the list as the unknown. This process continues until an answer is obtained, or until no more possible variables remain. In the latter case, PRESS fails to decide whether the expression is an identity or not.

In the example (xxv) above, PRESS applies Standard Homogenization, and rewrites the equation in terms of tan(x). Change of Unknown can then be applied, substituting y for tan(x). This produces the rational equation
$$(1 + y^2)/(1 - y^2) + 2 \cdot y/(1 - y^2) = (1 + y)/(1 - y).$$
Now Quotient Removal can be applied. The equation is multiplied by the product
$$(1 - y^2) \cdot (1 - y).$$
As discussed in section I.2.2, the multiplication by (1 - y) is not really necessary, the other term in the product contains this as a factor. However, PRESS does not check for this. The multiplication produces the polynomial
$$(1 + y^2) \cdot (1 - y) + 2 \cdot y(1 - y) = (1 + y) \cdot (1 - y^2).$$
On normal forming, this becomes 0 = 0. This statement has the value **true**, so PRESS concludes that the expression is an identity. Note that PRESS does not perform the back-substitution tan(x) = y.

The second part of the question above asks the student to solve an equation, either using the identity or not. The examiners expect the student to use the identity to replace the left hand side of the equation with the right hand side of the identity. The equation becomes
$$(1 + \tan(x))/(1 - \tan(x)) = 3,$$
which can be by easily solved by Change of Unknown, Quotient Removal, and Polynomial Methods.

At present, PRESS does not have a mechanism for using this identity to solve the given equation. PRESS solves it by first applying a trivial Change of Unknown, substituting y for 2·x. Standard Homogenization is then applied, in order to apply the major Change of Unknown. Quotient Removal and Polynomial methods are then used. This method is longer than the alternative above.

---

[12] Here we mean a variable in the equation solving sense, PRESS represents these variables using Prolog constants.

[13] Recall that the current implementation does not check for irreversible steps, so occasionally PRESS will return **true** when the expression is not an identity.

APPENDIX II

HOMOGENIZATION SPECIALISTS

This appendix concludes the description of the specialists used to choose the reduced term in Homogenization. The Simplicity Method and trigonometric case were described in Chapter 3.

II.1 Exponential Offenders Sets

In the case of the **Exponential Offenders Set** every term is of the form
$$u_i^{v_i}$$
where the u_i are constants and the v_i are expressions containing the unknown. There are two types of set that occur in the examples we have looked at. In both types the v_i are linear polynomials of the form
$$a_i \cdot x + b_i.$$
In the first type the u_i are the same for every term. The common value can be a rational, a non-rational constant such as e, or a symbolic constant, e.g. b.

In the second case the u_i can vary from term to term, but they must be rationals.[1]

An example of the first case is:
$$e^{3 \cdot x} - 2 \cdot e^x - 3 \cdot e^{-x} = 0 \quad \text{(London 1977)} \tag{i}$$
Here the offenders set is $\{e^x, e^{-x}, e^{3 \cdot x}\}$

An example of the second case is:
$$4^x - 2^{x+1} - 3 = 0 \quad \text{(A.E.B. 1971)} \tag{ii}$$

First note that when dealing with exponential offenders sets of either type if there is a term of the form u^{v+w} where u and w do not contain the unknown and v does, the term can be rewritten as $(u^w) \cdot u^v$. The first term of this product is some constant, so only the second term need now be rewritten as a function of the reduced term.

---

[1] Note that if all the u_i are the same rational, r say, then both types apply. In the case the first method is used as it involves less work.

II.1.1 The First Exponential Case

The terms in the set are all of the form u^{v_i}, u being a constant. All the v_i are of the form $a_i \cdot x + b_i$, where x is the unknown.[2] The method works by finding the greatest common divisor (g.c.d.) of the a_i's.[3]

As noted above the b_i may be disregarded for the purpose of finding the reduced term.

The reduced term is then $u^{g \cdot x}$, where g is the g.c.d of the a_i.

So, in (i) the a_i are 3, 1 & -1. The g.c.d is 1. (The g.c.d. is defined as positive except in the case where all the terms are negative). Therefore the reduced term is e^x.

An example of the rewrite rules involved is
$$U^V => (U^W)^{V/W}$$
Note that this rewrite rule has many apparently undesirable properties. It can loop and it decreases the simplicity of the equation. Homogenization can use these rules, because it applies them in a very directed way, towards a specific goal.

II.1.2 The Second Exponential Case

In the second case all the terms are of the form $(u_j)^{a_j \cdot x + b_j}$, where the u_j are rationals. The present implementation has another requirement that all the u_j are either less than one in magnitude, or that they are all greater. This restriction could be lifted, but none of our current examples require this.

If all the u_j are greater than one, the smallest u_j is found. Call this k.

If all the u_j are smaller than one, the largest u_j is found, call this k.

The method then checks if all the other u_j are powers of k, i.e. for every j there is a rational r_j, such that
$$(u_j)^{r_j} = k.$$
If this condition holds, the g.c.d. of the a_j, g, is found. The reduced term is then $k^{g \cdot x}$.

If the conditions are not satisfied, or g cannot be found, the simplicity method is tried.

So in (ii) the offenders set is $\{4^x, 2^{x+1}\}$. k is 2, g is 1. 4 is indeed a power of 2, so the reduced term is 2^x. Note that 2^{x+1} is rewritten as $2 \cdot 2^x$.

An example of the rewrite rules used is the following:
$$U^V => ((W^K)^{V/K})^M,$$

---

[2] In fact the method still applies if x is a term containing the unknown, provided it contains no multiplicative rational factor(which should be in a_i.) Thus x above can be replaced by tan(x), for example. However, for simplicity it will be assumed that x is just the unknown in the following discussion.

[3] To find the g.c.d of rationals express the rationals in terms of the lowest common denominator, k say, and take the g.c.d of the resulting numerators, h say. The g.c.d of the rationals is then h/k, reduced to the lowest terms.

where $U = W^M$. Note that this choice of reduced term will not always produce an output equation of the lowest degree. Consider the equation:

$$4^{2 \cdot x} + 4^x + 2^{2 \cdot x} = 24.$$

According to the algorithm above, the reduced term should be 2^x. Performing the rewrites and substituting $y = 2^x$, gives

$$y^4 + y^2 + y^2 = 24.$$

(Note that 4^x and $2^{2 \cdot x}$ are equal.) This is a quadratic in y^2, showing that a better choice of reduced term is $2^{2 \cdot x}$. The algorithm could be made more sophisticated to deal with cases such as this. Against this, the method minimizes the number of rewrite rules required, a rewrite rule to express 4^x in terms of $2^{2 \cdot x}$ is not needed. This economy in rewrite rules seems to be more important for efficiency than obtaining the lowest degree output equation.

II.2 The Logarithmic Case

PRESS uses a two argument predicate for log, $\log_x y$ being represented as $\log(x,y)$. To reduce the number of subscripts, this notation is adopted in this section.

The Homogenization method for the **Logarithmic Offenders Set** requires that every term in the offenders set must have at only one argument that contains the unknown. (It must have one, otherwise the term would not be in the offenders set). The other argument must be either a positive rational or a constant term.[4]

The offenders set is of the form

$$\{\log(n_1, f_1(x)), \ldots, \log(n_p, f_p(x)),$$
$$\log(f_{p+1}(x), n_{p+1}), \ldots, \log(f_q(x), n_q)\}$$

where the n_i are constant terms, and the $f_i(x)$ are functions containing the unknown.

There are four cases:

1. All the n_i are the same and so are the $f_i(x)$.

2. The n_i are all identical, but the $f_i(x)$ are not.

3. The $f_i(x)$ are all identical, but the n_i are not.

4. Neither the $f_i(x)$ nor the n_i are all identical.

II.2.1 The First Logarithmic Case

The first case, 1 above, is the simplest. In this case, the offenders set must be
$$\{\log(n, f(x)), \log(f(x), n)\}$$
(Both terms must occur since the offenders set must be non-singleton, otherwise Change of Unknown could be applied directly.) The term $\log(n, f(x))$ is chosen as the reduced term, and the rewrite rule

---

[4] A term free of the unknown, such as e or cos(a) where a is not the unknown.

$$\log(Y,X) \Longrightarrow 1/\log(X,Y)$$
is used. (Note that the choice of $\log(n,f(x))$ as the reduced term, rather than $\log(f(x),n)$ is entirely arbitrary.)

As an example, the equation
$$4 \cdot \log_x 2 + \log_2 x = 5 \qquad \text{(London 1978)}$$
gives rise to this type of offenders set, $\log_2 x$ is chosen as the reduced term.

II.2.2 The Second Logarithmic Case

The implementation can only deal with a restricted version of case 2 at present. All the $f_i(x)$ must be of the form x^{r_i} where the r_i are rationals. In this case the reduced term is[5] $\log(n,x^g)$, where n is the common value of the n_i, and g is the greatest common divisor of the r_i.[6]

An example of an equation giving rise to this type of offenders set is:
$$\log(e,x^2) + \log(x,e) = 1.$$
The reduced term in this case is $\log(x,e)$. The rewrite rule used is
$$\log(A,X^N) \Longrightarrow N \cdot (\log(X,A))^{-1}.$$

II.2.3 The Third Logarithmic Case

For case 3, there are two choices of methods, the **power method**, and the **base ten method**. The power method will be discussed first.

II.2.3.1 The Power Method

For the **power method** to apply, all the n_i must be positive rationals. If this requirement is met, the reduced term is selected as follows:

- If all the n_i are greater than one then the smallest of the n_i is found, n_s say.

- If all the n_i are less than one then the largest is found, n_s say.

- If neither of the above hold the method fails and the base ten method is tried instead, see below.

- It is then checked that all the n_i are powers of n_s. If this is not the case the base ten method is tried.

- The reduced term is then $\log(n_s,f(x))$ or $\log(f(x),n_s)$, depending on which of these actually appears in the offenders set. If both occur (as in the original example above) then select $\log(n_s,f(x))$.

An example of the rewrite rules used is

---

[5] Alternatively, the reduced term could be $\log(x^g,n)$. The reduced term $\log(x^g,n)$ is chosen if the equation contains it, otherwise $\log(n,x^g)$ is used. Again, this is an arbitrary decision.

[6] The definition of the greatest common divisor of rationals was given above.

$\log(U^N,V) \Longrightarrow (N \cdot \log(V,U))^{-1}$.

II.2.3.2 The Base 10 Method

If the power method fails, the **base ten method** is used. In this case the reduced term is $\log(10,f(x))$. The rewrite rules used are

$\log(A,X) \Longrightarrow (\log(10,A))^{-1} \cdot \log(10,X)$

$\log(X,A) \Longrightarrow \log(10,A) \cdot (\log(10,X))^{-1}$

where A does not contain the unknown, thus $\log(10,A)$ is a constant.

Why use base 10? Any other base would do, but students taking A level exams are provided with base 10 log tables. For this reason, the answers provided by the board refer to base 10. Thus, by using base 10, the answers of PRESS can be compared with the given answers easily. Also, this choice makes PRESS output resemble human output slightly more than any other base.

As an example of an equation for which this method is needed consider:
$\log(x,2) \cdot \log(x,3) = 5$ (London Board 1981)
Homogenization transforms this to
$\log(10,2) \cdot \log(10,x) \cdot \log(10,3) \cdot \log(10,x) = 5$ (iii)
which is easily solved after Change of Unknown.[7]

II.2.4 The Fourth Logarithmic Case

Case 4 combines the features of the other cases. The $f_i(x)$ must be as in case 2, and the n_i are as in case 3. The reduced term will be one of the set

$\log(10,x^g), \log(n_s,x^g), \log(x^g,n_s)$

(where x^g is defined in case 3 and n_s in case 2.)

II.3 The Hyperbolic Offenders Set

The **Hyperbolic Offenders Set** method is applicable to offenders sets of the type where all terms in the set are hyperbolic functions with arguments $r_i \cdot x$, where x is the unknown and r_i is a rational.

The first step is similar to that of the trigonometric case. The offenders set is examined to see if it is a sinh-cosh, sech-tanh, or cosech-coth pair, where these are the obvious analogues of the trigonometric case. Again, if the set is one of these types, and one of the pair occurs only to an even power, the other member of the pair is chosen as the reduced term. An example of this type of equation is:

$3 \cdot \text{sech}^2(x) + 4 \cdot \tanh(x) + 1 = 0$ (A.E.B. 1971)

Normally, these conditions are not satisfied and another method is used. In this case the g.c.d of the r_i is found, g say. The reduced term is then $e^{g \cdot x}$. The hyperbolic functions are then written in terms of $e^{g \cdot x}$ by using the standard definitions. Thus if the offenders set were $\{\cosh(x), \sinh(2 \cdot x)\}$ the reduced term would be e^x, so $\cosh(x)$ would be rewritten as

---

[7] Alternatively, equation (iii) can be solved by Collection followed by Isolation.

$(e^x + (e^x)^{-1})/2$,
and $\sinh(2 \cdot x)$ becomes
$((e^x)^2 - (e^x)^{-2})/2$.

Most of the hyperbolic equation questions we have found on A level papers were easily dealt with. The following is typical.
$5 \cdot \cosh(x) - 3 \cdot \sinh(x) = 5$ (London June 1978)

II.4 The Exponential and Hyperbolic Offenders Set

The **Exponential and Hyperbolic Offenders Set** case is a mixture of the hyperbolic and first exponential cases. The hyperbolic and exponential are treated separately. A reduced term is selected from the $e^{f(x)}$ terms, using exactly the same process as described in section II.1.1. Suppose that the reduced term is $e^{a \cdot x}$.

Another reduced term is selected from the hyperbolic set, using a similar process to that described in the previous section, except that the reduced term must always be of the form $e^{g(x)}$, as the sinh-cosh pairs etc cases are not applicable. Call this reduced term $e^{b \cdot x}$.

The final reduced term is $e^{g \cdot x}$ where g is the g.c.d of a and b.

For example, consider the equation
$\sinh(x) - 3 \cdot e^x + e^{3 \cdot x} = 0$.
The offenders set is
$\{\sinh(x), e^x, e^{3 \cdot x}\}$,
the set is divided into the hyperbolic set
$\{\sinh(x)\}$
and the exponential set
$\{e^x, e^{3 \cdot x}\}$.
The reduced term from the hyperbolic set is e^x. The reduced term from the exponential set is also e^x. Therefore, the reduced term from the whole set is e^x.

The output equation is
$(e^x + (e^x)^{-1})/2 - 3 \cdot e^x + (e^x)^3 = 0$.
After Change of Unknown and some polynomial manipulation, this becomes the disguised quadratic
$2 \cdot y^4 - 5 \cdot y^2 + 1 = 0$,
where y has been substituted for e^x.

APPENDIX III

GENERATING NEW PROBLEMS

Like LEX2, [55, 56], LP is able to create new problems for it to solve, using its **problem generator**. LP uses the new problems to test the general applicability of the newly created schema. After LP has completed its examination of the worked example, it produces a set of equations that differ in only one feature from the generating equation. The simplest change is the replacing of one numerical coefficient by another. Another consists of replacing a function by a sibling function, e.g. replacing sin by cos.

For example, if the program has just produced a schema for the equation
$\cos(x) + 2 \cdot \cos(2 \cdot x) + \cos(3 \cdot x) = 0$,
the problem generator generates the following new equations:

1) $\quad 2 \cdot \cos(x) + 4 \cdot \cos(2 \cdot x) + \cos(3 \cdot x) = 0$

2) $\quad \cos(x) + 4 \cdot \cos(2 \cdot x) + 2 \cdot \cos(3 \cdot x) = 0$

3) $\quad \cos(x) + 2 \cdot \cos(2 \cdot x) + 2 \cdot \cos(3 \cdot x) = 0$

4) $\quad \cos(x) + 2 \cdot \cos(2 \cdot x) + \cos(4 \cdot x) = 0$

5) $\quad \sin(x) + 2 \cdot \cos(2 \cdot x) + \cos(3 \cdot x) = 0$

6) $\quad \cos(x) + 2 \cdot \cos(2 \cdot x) + \sin(3 \cdot x) = 0$

7) $\quad \cos(x) + 2 \cdot \cos(3 \cdot x) + \cos(3 \cdot x) = 0$

8) $\quad \cos(x) + 2 \cdot \sin(2 \cdot x) + \cos(3 \cdot x) = 0$

9) $\quad \cos(x) + 3 \cdot \cos(2 \cdot x) + \cos(3 \cdot x) = 0$

The right-hand side is not varied as the schema method is of type **Factorization**, and LP assumes that the right-hand side must be 0. Note that the present system does not vary the non-constant arguments of the trigonometric terms, nor the + functions.

The user selects which of these problems should be attempted. LP then attempts to use the schema to solve the new equation. At present, LP simply notes if the schema allowed it to solve the test problem or not.

In a more general self-improving algebra system, the results of the test would allow the program to refine the schema or the matcher appropriately. For example, only the last of the equations above can be solved by following the schema exactly. Such a result would be useful for a concept learning system, as it suggests that the coefficient of the middle cosine time can be generalized to any integer if the rest of the equation remains the same. The information could be passed to the fuzzy matcher, or it could modify information within the schema. Also, the fact that the schema cannot be used on the rest of the problems, goes some way towards discovering the class of equations for which the schema works exactly.[1]

LEX, [55, 56], attempts to generate only those test problems that it knows will provide helpful information to its refinement system. In contrast, the problem generator of LP has no such considerations. It performs simple syntactic peturbations of the generating equation. It appears to work well on the example above simply because the target class is so precise, i.e. small syntactic changes can move the equation outside this class.

It may seem desirable to have a problem generator that suggests only "helpful" problems. However, there is a danger in such an approach. In general, the problem generator may have to possess a considerable amount of the knowledge that the rest of the program is attempting to learn![2] For this reason, we prefer to have a "stupid" problem generator.

Mitchell, in [54], reports that he is dissatisfied with the problem generator of LEX2, as most of the "helpful" examples it produces cannot be solved. This is the case with LP's problem generator as well, and is to be expected as the target concept is very sharp. Mitchell suggests that the problem generator of LEX2 may have to be redesigned to overcome this. Presumably this would involve giving the generator extra knowledge. This seems to fall into the trap mentioned in the previous paragraph, but Mitchell does not discuss this point.

---

[1] The class consists of equations of the form
$a \cdot \cos(p) + b \cdot \cos(q) + a \cdot \cos(r) = 0$,
where $q = (p + r)/2$. This class is of course a subset of that solved by the Arithmetic Progression specialist of PRESS, (section 2.3.1.9).

[2] LEX does not in fact suffer from this problem, its problem generator needs no special knowledge. Usually the concepts that LEX is trying to learn can easily be expressed in the description language and the version space approach used by LEX allows the program to find the interesting new examples in such cases. If the situation is more complex, the problem generator will not be so helpful.

APPENDIX IV

PRESS OUTPUT

This appendix shows four annotated examples of output produced by the PRESS program. The output has not been changed in any way, except for the addition of comments, the use of italics, and the fact that some lines have been broken in two so that they fit on the page. Space considerations restrict the number of examples that can be included, so not every PRESS method is demonstrated. However, the examples included are fairly typical, and should give some idea of the output produced by PRESS.

All annotations are preceded by the " % " symbol, user input is in *italics*.

PRESS   (13 Jul 83)

```
                % The first example shows PRESS
                % using Homogenization on a
                % single equation.
```
| ?- *solve(log(2,x) + 4*log(x,2) = 5)*.

```
                % This problem comes from the
                % London 1978 A level exam.
                % The | ?- symbol is the top
                % level prompt of Prolog.
                % PRESS uses a two argument
                % log function, the first
                % argument is the base.
                % PRESS solves for x unless
                % told  otherwise.
```
Solving log(2, x) + 4 * log(x, 2) = 5 for x

Rewriting equation in terms of log(2, x)
gives log(2, x) + 4 * log(2, x) ^ -1 = 5

```
                % PRESS uses Homogenization here,
                % rewriting log(x,2) as 1/log(2,x).
                % ^ is the exponentiation operator.
```

Substituting   x1 for log(2, x) gives
4 * x1 ^ -1 + x1 = 5

> % Change of Unknown. The resulting
> % changed variable equation
> % is almost a polynomial in x1.

Multiply through by x1 to get

x1 ^ 2 + -5 * x1 + 4 = 0

> % PRESS converts the expression
> % into a proper polynomial
> % using the Polynomial methods.

Using quadratic equation formula

Solutions are x1 = 4 and x1 = 1

> % As it says! PRESS has found
> % the two solutions of the
> % changed variable equation.
> % Now it forms the
> % substitution equations.

Applying substitution
    x1 = log(2, x)

   to   :
    x1 = 4 # x1 = 1

> % # is used by PRESS to
> % indicate disjunction.

   gives :
    log(2, x) = 4 # log(2, x) = 1

> % Solving the first substitution equation.

Solving disjunct log(2, x) = 4

    x = 16
        (by Isolation)

> % Isolation is used to find the solution.
>
> % Solving the other substitution equation.

Solving disjunct log(2, x) = 1

    x = 2
        (by Isolation)

Answer is :
(X1 # X2)
  where :
    X1 = x = 16
    X2 = x = 2

yes
                        % PRESS has found the solutions of the
                        % original equation, so the task is
                        % finished.

                        % The next example shows
                        % Extended Homogenization

PRESS   (11 Feb 84)

| ?- $sim(cos(x)+cos(y)=1 \ \& \ sec(x) + sec(y) = 4)$.
                        % This equation comes from the 1976
                        % A.E.B. Paper
Simultaneously solving :
    cos(y) + cos(x) = 1
    sec(y) + sec(x) = 4
 For [x, y].
Homogenizing equations in x
 gives
    cos(y) + cos(x) = 1
    sec(y) + cos(x) ^ -1 = 4
                        % The equations are now homogeneous
                        % in x, so Change of Unknown
                        % can be applied.
Substituting x1 = cos(x) gives
    x1 + cos(y) = 1
    x1 ^ -1 + sec(y) = 4
Homogenizing equations in y
 gives
    x1 + cos(y) = 1
    x1 ^ -1 + cos(y) ^ -1 = 4

                    % Similarly for y
Substituting x2 = cos(y) gives
    x2 + x1 = 1
    x2 ^ -1 + x1 ^ -1 = 4
Solving x1 + x2 = 1 for x1
                        % PRESS picks the first equation
                        % as its linear in x1 and x2.
Tidying to x1 = 1 + -1 * x2

    x1 = 1 + -1 * x2
        (by Isolation)

Answer is :
X1
  where :
    X1 = x1 = 1 + -1 * x2

Applying substitution
$$x_1 = 1 + -1 * x_2$$

to   :
$$x_1 \wedge -1 + x_2 \wedge -1 = 4$$

gives :
$$x_2 \wedge -1 + (1 + -1 * x_2) \wedge -1 = 4$$

Solving $x_2 \wedge -1 + (1 + -1 * x_2) \wedge -1 = 4$ for x_2

Clearing of rational functions

$$x_2 \wedge -1 * (1 + -1 * x_2) = 3 + -4 * x_2$$
    % Using Quotient Removal. Note
    % that $x_2 \wedge -1$ has not been removed
    % as the Polynomial Methods do this.

Polynomial $-3 + 4 * x_2 + (1 + -1 * x_2) * x_2 \wedge -1$ becomes

$4 * x_2 + -4 + x_2 \wedge -1$ when in normal form

Multiply through by x_2 to get

$$4 * x_2 \wedge 2 + -4 * x_2 + 1 = 0$$

Using quadratic equation formula

The discriminant is zero, so the single solution is $x_2 = (1/2)$

Answer is :
X1
  where :
    $X_1 = x_2 = (1/2)$

Substituting back in x1 solution

Applying substitution
$$x_2 = (1/2)$$

to   :
$$x_1 = 1 + -1 * x_2$$

gives :
$$x_1 = (1/2)$$

Applying substitution
$$x_2 = \cos(y)$$

to   :
$$x_2 = (1/2)$$

gives :
$$\cos(y) = (1/2)$$

    Letting n1 denote an arbitrary integer
    $y = 60 + 360 * n_1$ # $y = -60 + 360 * n_1$
    (by Isolation)

```
Applying substitution
    x1 = cos(x)

   to   :
    x1 = (1/2)

   gives :
    cos(x) = (1/2)

        Letting n2 denote an arbitrary integer
    x = 60 + 360 * n2 # x = -60 + 360 * n2
        (by Isolation)


Final Answers are :
((X1 # X2) & (X3 # X4))
  where :
    X1 =  y = 60 + 360 * n1
    X2 =  y = -60 + 360 * n1
    X3 =  x = 60 + 360 * n2
    X4 =  x = -60 + 360 * n2

                % The final answer.
yes
                % The next problem is from the
                % A.E.B. 1971 A level.  Appendix
                % V shows how LP
                % solves this problem.
```

| ?- $solve(cos(x) + 2{*}cos(2{*}x) + cos(3{*}x) = 0)$.

```
Solving cos(x) + 2 * cos(2 * x) + cos(3 * x) = 0 for x

Angles are in arithmetic progression
                % This condition is discovered
                % by the arithmetic progression
                % trig specialist.  This message
                % is printed out when the angles
                % have passed the test.  The method
                % may nevertheless not be applicable,
                % see the example in section 2.3.1.9.
Adding in pairs
2 * cos(2 * x) + 2 * cos(2 * x) * cos(x) = 0

(2 + 2 * cos(x)) * cos(2 * x) = 0

                % In this case, the method does apply, and
                % the equation is split into factors.
```

```
                    % Solving the first factor.

Solving factor cos(2 * x) = 0

        Letting n1 denote an arbitrary integer
   x = 45 + (1/2) * (180 * n1)
        (by Isolation)

                    % The solution is obtained by Isolation.
                    % Now for the other factor.
Solving factor 2 + 2 * cos(x) = 0

        Letting n2 denote an arbitrary integer
   x = 180 + 360 * n2 # x = -180 + 360 * n2
        (by Isolation)

                    % This time Isolation returns
                    % a disjunctive solution.
Answer is :
( X1 # ( X2 # X3 ) )
  where :
    X1 =   x = 45 + 90 * n1
    X2 =   x = 180 + 360 * n2
    X3 =   x = -180 + 360 * n2

                    % The final solution.

yes

| ?-solve(log(e,x+1) + log(e,x-1) = 3).
```

$$\text{\% This equation is the standard}$$
$$\text{\% example in [21].}$$

```
Solving log(e, x + 1) + log(e, x - 1) = 3 for x

Tidying to log(e, 1 + x) + log(e, -1 + x) = 3
log(e, (-1 + x) * (1 + x)) = 3

                    % Attraction
log(e, -1 + x ^ 2) = 3

                    % Collection
     x = (1 + e ^ 3) ^ (1/2) # x = -1 * (1 + e ^ 3) ^ (1/2)
        (by Isolation)

                    % Isolation
```

```
Answer is :
( X1 # X2 )
  where :
    X1 =   x = (1 + e ^ 3) ^ (1/2)
    X2 =   x = -1 * (1 + e ^ 3) ^ (1/2)
                % The final answer.
                % Note that X2 is not really an
                % acceptable solution, as
                % it makes the arguments of the logs
                % negative.  PRESS does not vet the
                % solutions at present.

yes
| ?- halt.
                % End the run
```

APPENDIX V

LP OUTPUT

This appendix contains examples of LP's output. As in the previous appendix, the output has not been changed in any way, except for the addition of comments, the adjustment of line widths and the *italicising* of user input. Annotations are preceded by the "%" symbol.

```
Learning PRESS   Mark 4 (29 Feb 84)

Type ' help. <CR> ' for help

[User [400, 4322]. Session begins at 18:22:27 on 1.3.1984]
 | ?- disable cifer.

Your terminal is not a cifer.
                % Prevents the program outputting
                % clear screen control sequences.
yes
                % The first example is very simple,
                % teaching LP a new Isolation rule
                % involving a new made up function.

 | ?- give_example.
                % Use inputs a worked example.

Enter example, line by line, using x as the unknown.

Terminate with '<CR>' on new line.
```

First line: $f(x) = 1$.

```
f(x) = 1.
                % LP outputs the tidied form of
                % the line.
```

Next line: $x = g(1)$.

```
x = g(1).
```

Next line:

End of Example Input.

Example has been stored, and can be rerun by typing xredo.

[Run begins at 18:23:10]

```
[**Warning, LP has no rules for functor f/1**]
                % LP warns the user that it does not know
                % about the function f.

[Processing]    % LP is working on operator identification.
                % LP only outputs the results when this stage
                % is complete.

        Operator Identification Complete.
```

Solving
```
                f(x) = 1
```
for x

The steps were as follows.

Step not understood, continuing processing

```
        [Possibly missing rule for method Isolation.]

                % LP notes that the step satisfies the
                % preconditions and postconditions of
                % Isolation, but Isolation does not apply.
                % Maybe a rule is missing...
x = g(1).
```

[Solution]

```
        End of Example.
```

Trying to conjecture reasons for unknown step.

LP OUTPUT

```
Working on step from

f(x) = 1

to

x = g(1)

Step is an application of Isolation


Conjecture that

f(x) = 1

                    ->

x = g(1)
```
                % The only possible conjecture
                % but the user is asked to confirm.

```
Is the conjecture correct? (y/n)
Confirm:
```  *y*

```
Enter the rewrite rule, in the form Old => New,
terminate with a "."
(You will then be prompted for further details)

Enter Rule
Rule:
```  $f(A){=}B \implies A{=}g(B)$.
                % The general (made up) rule!
```
Enter Unknown
Terminate with <cr>
(Just type <cr> if no distinguished unknown.)
Unknown:
[No distinguished unknown]
```
                % Note that A need not be the unknown
                % The new Isolation rule is valid
                % as long as A contains the unknown.
                % We do not have to supply this condition
                % as it is a condition of the parent methods
                % (Isolation and Nasty Function Methods)
```
Enter List of conditions
Terminate with <cr> (Just type <cr> if none.)
Conditions:

[No conditions]

Rule is being stored as a Isolation rule.
```
                % The step was marked as an
                % application of Isolation,
                % so the new rule is stored as an
                % Isolation rule.

APPENDIX V

[Adding functor f/1 to list of known functors]

    % Now LP knows one rule about f!

End of Conjecture.

[Creating new schema method, called auto1, for this equation]
    % The (rather trivial) schema is shown below.

[Example took 926 milliseconds]
    % Timing information on a Dec-10 KL.

Do you want to run the problem again now? Type y or n.
Reply:y

Rerunning example.

[Run begins at 18:23:25]

[Processing]

  Operator Identification Complete.

Solving
    f(x) = 1
for x

The steps were as follows.

Applying Isolation
    % Now step is understood

 x = g(1).

[Solution]

  End of Example.

No difficulties encountered.

[Already have this schema for this equation]

[Example took 286 milliseconds]

[Type generate_problem to test the schema]

[End of Output]

yes

    % Now LP is given an equation,
    % using the new function.

| ?- $solve(f(x) - 3*f(x) + 2 = 0)$.

[Run begins at 18:23:33]

Solving f(x) - 3 * f(x) + 2 = 0 for x.

Tidying to -3 * f(x) + f(x) = -2.

Applying substitution x1 = f(x) to obtain
x1 + -3 * x1 = -2
                    % Change of Unknown

Using Polynomial methods to obtain

x1 = 1.
                    % Changed variable equation
                    % is solved.
Applying substitution f(x) = x1 to obtain
f(x) = 1

Attempting to use schema method method(auto1),
generated for equation

f(x) = 1.
                    % Now LP finds the schema created above!
Trying indicated schema step of applying Isolation.

Isolating Equation to obtain

x = g(1).

[End of solution]

Answer is : x = g(1)

[Type work_solution to create a schema for this equation]

[Problem took 750 milliseconds]

yes
                    % The next example is the standard one used
                    % throughout chapter 4.
                    % Appendix IV shows how PRESS
                    % solves this problem.
| ?- *wep*.
                    % Read in a file containing worked examples.

learn:wep consulted   2229 words   0.92 sec.

yes
| ?- *enable output*.

The example will now be output.
                    % LP now prints out the example before
                    % processing it. (Normally, the example
                    % is not output, as the user knows what
                    % it is, and the annotated example
                    % is output later anyway.)
yes

APPENDIX V

```
| ?- trig1.
                    % The name of the example.
[Run begins at 18:23:58]
Working on the following example, x is the unknown.

        cos(x) + 2 * cos(2 * x) + cos(3 * x) = 0.

        2 * cos(x * 2) * cos(x) + 2 * cos(x * 2) = 0.

        2 * cos(x * 2) * (cos(x) + 1) = 0.

        cos(2 * x) = 0   v   cos(x) + 1 = 0.

        cos(2 * x) = 0.

        x = 180 * n1 * (1 / 2) + 45.

        cos(x) + 1 = 0.

        x = n2 * 360 + 180.

        [End of Worked Example]
[Processing]
                % LP prints out this message
[Processing]    % at various stages in the analysis,
                % the longer the analysis, the
[Processing]    % more times this message appears.
                % It is there to reassure the user
[Processing]    % of a heavily loaded system that
                % the computer has not crashed!

        Operator Identification Complete.


Solving
            cos(x) + 2 * cos(2 * x) + cos(3 * x) = 0
for x
The steps were as follows.

Step not understood, continuing processing
                    % LP does not have the required rule

  2 * cos(2 * x) * cos(x) + 2 * cos(2 * x) = 0.

Applying Prepare for Factorization

  2 * cos(2 * x) * (1 + cos(x)) = 0.

Applying Factorization

    cos(2 * x) = 0   v   cos(x) = -1.
```

Solving first factor

 $\cos(2 * x) = 0$.

Applying Isolation

 $x = 45 + 90 * n1$.

[Solution]

Solving next factor

 $\cos(x) = -1$.

Applying Isolation

 $x = 180 + 360 * n2$.

[Solution]

      End of Example.

Trying to conjecture reasons for unknown step.

Working on step from

$\cos(x) + 2 * \cos(2 * x) + \cos(3 * x) = 0$

to

$2 * \cos(2 * x) * \cos(x) + 2 * \cos(2 * x) = 0$

Conjecture that

$\cos(3 * x) + \cos(x)$

          =

$2 * \cos(2 * x) * \cos(x)$.

Is the conjecture correct? (y/n)
Confirm: y

Enter the rewrite rule, in the form Old => New, terminate with a "."
(You will then be prompted for further details)

Enter Rule
Rule: $cos(A)+cos(B) => 2*cos((A+B)/2)*cos((A-B)/2)$.

Enter Unknown
Terminate with <cr>
(Just type <cr> if no distinguished unknown.)
Unknown:

[No distinguished unknown]

```
Enter List of conditions
Terminate with <cr> (Just type <cr> if none.)
Conditions:
```

[No conditions]

```
Give a name for this rule (use an atom).
Name:
```
cos_sum.

End of Conjecture.

Trying to find a reason why the step from

$$\cos(x) + 2 * \cos(2 * x) + \cos(3 * x) = 0$$

to

$$2 * \cos(2 * x) * \cos(x) + 2 * \cos(2 * x) = 0$$

was performed.

Reason found is that next method to be applied,

>Prepare for Factorization

requires its input to satisfy the following unsatisfied precondition:

Precondition	Explanation
common_subterms/3	Equation has common additive subterms

% LP now creates a new method

Creating a new method, named method(auto2),

that transforms equation so that the following precondition is satisfied:

<common_subterms/3>

[Creating new schema method, called auto3, for this equation]

[Example took 3674 milliseconds]

Do you want to run the problem again now? Type y or n.
Reply:*y*

Rerunning example.

[Run begins at 18:24:41]

Working on the following example, x is the unknown.

$$\cos(x) + 2 * \cos(2 * x) + \cos(3 * x) = 0.$$

$$2 * \cos(x * 2) * \cos(x) + 2 * \cos(x * 2) = 0.$$

$$2 * \cos(x * 2) * (\cos(x) + 1) = 0.$$

$$\cos(2 * x) = 0 \quad v \quad \cos(x) + 1 = 0.$$

$$\cos(2 * x) = 0.$$

$$x = 180 * n1 * (1 / 2) + 45.$$

$$\cos(x) + 1 = 0.$$

$$x = n2 * 360 + 180.$$

[End of Worked Example]

[Processing]

[Processing]

[Processing]

[Processing]

Operator Identification Complete.

Solving
$$\cos(x) + 2 * \cos(2 * x) + \cos(3 * x) = 0$$
for x

The steps were as follows.

Applying method(auto2)
% This step is now understood.

$$2 * \cos(2 * x) * \cos(x) + 2 * \cos(2 * x) = 0.$$

Applying Prepare for Factorization

$$2 * \cos(2 * x) * (1 + \cos(x)) = 0.$$

Applying Factorization

$$\cos(2 * x) = 0 \quad v \quad \cos(x) = -1.$$

APPENDIX V

Solving first factor

  $\cos(2 * x) = 0$.

Applying Isolation

  $x = 45 + 90 * n1$.

[Solution]
Solving next factor

  $\cos(x) = -1$.

Applying Isolation

  $x = 180 + 360 * n2$.

[Solution]

        End of Example.

No difficulties encountered.

[Already have this schema for this equation]
            % No need to duplicate the schema.

[Example took 2637 milliseconds]

[Type generate_problem to test the schema]

[End of Output]

yes

            % Now LP is given a similar equation
            % to solve.

| ?- $solve(2*cos(x) + 3*cos(3*x) + 2*cos(5*x) = 0)$.

[Run begins at 18:24:50]

Solving $2 * \cos(x) + 3 * \cos(3 * x) + 2 * \cos(5 * x) = 0$
for x.

Attempting to use schema method method(auto3),
generated for equation

$\cos(x) + 2 * \cos(2 * x) + \cos(3 * x) = 0$.

            % LP tries to use the new schema.

Attempting to manipulate equations into factors.
            % Schema is of type Factorization.

LP OUTPUT

Trying indicated schema step of applying method(auto2).

Failed to apply schema step method(auto2).

> % auto2 cannot be applied
> % straight away.

Trying to find a method whose result satisfies the following precondition

Precondition	Explanation
common_subterms/3	Equation has common additive subterms

while keeping the following conditions true

Precondition	Explanation
rhs_zero/1	Right hand side of equation is 0
is_sum/2	Left hand side of equation is a sum

[Method Attraction can be applied
but fails to satisfy (at least) the precondition

Precondition	Explanation
common_subterms/3	Equation has common additive subterms

]

> % The Table of Connections shows that
> % a number of methods might be useful.
> % However, only Attraction can
> % actually be applied. However,
> % Attraction in fact fails to
> % satisfy at least one of the
> % conditions, common_subterms/3.

Can't find method to help satisfy new conditions.

> % So it looks for a method that
> % does not undo anything already achieved.

Looking for method that keeps satisfied conditions satisfied.

[Method Attraction can be applied
and satisfies all the conditions]

> % Attraction is OK for this purpose

Attracting Equation to obtain

$3 * \cos(3 * x) + 2 * (\cos(x) + \cos(5 * x)) = 0.$

Trying indicated schema step of applying method(auto2).

Using rule cos_sum to apply new method,
auto2, to obtain
                % Now the schema method can be used
                % without modification.

cos(x * 3) * 3 + cos(x * 2) * (cos(x * 3) * 4) = 0.

Trying indicated schema step of applying
Prepare for Factorization.

Preparing for Factorization to obtain

(3 + 4 * cos(2 * x)) * cos(3 * x) = 0.

Solving cos(3 * x) = 0 for x.

Choosing schema to solve this factor.

Trying indicated schema step of applying Isolation.

Isolating Equation to obtain

x = 30 + 60 * n5.

[End of solution]

Solving 3 + 4 * cos(2 * x) = 0 for x.

Choosing schema to solve this factor.

Trying indicated schema step of applying Isolation.

Isolating Equation to obtain

x = (1/2) * (360 * n6 + arccos((-3/4)))   v
x = (1/2) * (360 * n6 + -1 * arccos((-3/4))).

[End of solution]

Answer is : x = 30 + 60 * n5   v
             x = (1/2) * (360 * n6 + arccos((-3/4)))   v
             x = (1/2) * (360 * n6 + -1 * arccos((-3/4)))

[No problems with schema]

[Problem took 4377 milliseconds]

yes

                % The next example shows LP learning
                % a instance of Collection.

LP OUTPUT

| ?- *disable collection*.

Method Collection turned off.

     % Now LP is unable to use Collection.

yes

| ?- *disable output*.
     % This feature is no longer needed.

The example will now not be output.

yes

| ?- *give_example*.

Enter example, line by line, using x as the unknown.

Terminate with '<CR>' on new line.

First line: *log(e,x + 1) + log(e, x - 1) = 3*.

log(e, -1 + x) + log(e, 1 + x) = 3.

Next line: *log(e, (x+1)*(x-1)) = 3*.

log(e, (-1 + x) * (1 + x)) = 3.

Next line: *log(e, x^2 - 1) = 3*.

log(e, -1 + x ^ 2) = 3.

Next line: $x = (e\char`^3 + 1)\char`^(1/2) \# x = -1*(e\char`^3 + 1)\char`^(1/2)$.

x = (1 + e ^ 3) ^ (1/2)   v x = -1 * (1 + e ^ 3) ^ (1/2).

     % LP converts # into v.

Next line:

End of Example Input.

Example has been stored, and can be rerun typing xredo.

[Run begins at 18:25:42]

[Processing]

[Processing]

[Processing]

Operator Identification Complete.

Solving
$$\log(e, -1 + x) + \log(e, 1 + x) = 3$$
for x

The steps were as follows.

Applying Attraction

$\log(e, (1 + x) * (-1 + x)) = 3.$

Step not understood, continuing processing

$\log(e, -1 + x \char`\^\ 2) = 3.$

% The Collection Step

Applying Isolation

$x = (1 + e \char`\^\ 3) \char`\^\ (1/2) \quad v \quad x = -1 * (1 + e \char`\^\ 3) \char`\^\ (1/2).$

[Solution]

End of Example.

Trying to conjecture reasons for unknown step.

Working on step from

$\log(e, (1 + x) * (-1 + x)) = 3$

to

$\log(e, -1 + x \char`\^\ 2) = 3$

Conjecture that

$(1 + x) * (-1 + x)$

=

$-1 + x \char`\^\ 2.$

Is the conjecture correct? (y/n)
Confirm:*y*

Enter the rewrite rule, in the form Old => New,
terminate with a "."
(You will then be prompted for further details)

Enter Rule
Rule:*(A+B)*(A-B) => A^2 - B^2.*

```
Enter Unknown
Terminate with <cr>
(Just type <cr> if no distinguished unknown.)
Unknown:

[No distinguished unknown]

Enter List of conditions
Terminate with <cr> (Just type <cr> if none.)
Conditions:

[No conditions]

Give a name for this rule (use an atom).
Name:
```
collect_rule.

End of Conjecture.

Trying to find a reason why the step from

$$\log(e, (1 + x) * (-1 + x)) = 3$$

to

$$\log(e, -1 + x \char`\^ 2) = 3$$

was performed.

Reason found is that next method to be applied,

Isolation

requires its input to satisfy the following unsatisfied precondition:

Precondition	Explanation
single_occ/2	Unknown occurs exactly once in equation

Creating a new method, named method(auto4),

that transforms equation so that the
following precondition is satisfied:

<single_occ/2>

APPENDIX V

```
% Note that auto4 is not quite Collection,
% for three reasons:
% 1) Obviously, at present it only has
% one rule, but more can be learned later.
% 2) More importantly it differs from
% Collection in that it insists that
% Isolation must be applicable after
% its application, Collection can be
% applied in other circumstances.
% 3)  It can apply to non least-dominating
% terms.
```

[Creating new schema method, called auto5, for this equation]

[Example took 2750 milliseconds]
Do you want to run the problem again now? Type y or n.
Reply:*y*

Rerunning example.

[Run begins at 18:26:33]

[Processing]

[Processing]

[Processing]

   Operator Identification Complete.

Solving
$$\log(e, -1 + x) + \log(e, 1 + x) = 3$$
for x

The steps were as follows.

Applying Attraction

$\log(e, (1 + x) * (-1 + x)) = 3.$

Applying method(auto4)

$\log(e, -1 + x \char`\^ 2) = 3.$

Applying Isolation

$x = (1 + e \char`\^ 3) \char`\^ (1/2) \quad v \quad x = -1 * (1 + e \char`\^ 3) \char`\^ (1/2).$

[Solution]

   End of Example.

No difficulties encountered.

[Already have this schema for this equation]

```
[Example took 1829 milliseconds]

[Type generate_problem to test the schema]

[End of Output]

yes
| ?- show schemas.
                    % LP shows the some of the important
                    % parts of the schemas.  (To avoid
                    % swamping the user with output,
                    % it does not show the already
                    % satisfied preconditions)

                    Schema Method method(auto5)


Method                              Purpose

Attraction                          <none>

auto4                               <single_occ/2>

Isolation                           <key method>
_____

[End of Schema]

Type:General

Generating Equation:log(e,-1+x)+log(e,1+x)=3

Unknown:x
                    % The purpose of the Attraction Step
                    %  is outside the description space
                    %  used by LP, hence the <none> in
                    %  the first entry.
```

Schema Method method(auto3)

Method	Purpose
auto2	<common_subterms/3>
Prepare for Factorization	<is_product/2>
Factorization	<key method>

Isolation	<key method>

Isolation	<key method>

[End of Schema]

Type:Factorization

Generating Equation:cos(x)+2*cos(2*x)+cos(3*x)=0

Unknown:x

Schema Method method(auto1)

Method	Purpose
Isolation	<key method>

[End of Schema]

Type:General

Generating Equation:f(x)=1

Unknown:x

yes

```
                % Now a demonstration of the
                % problem generator, see
                % appendix III.
| ?- generate_problem.
```

Attempting to generate a problem to test schema generated by problem
$\log(e, -1 + x) + \log(e, 1 + x) = 3$

The set of generated problems is:

|1) $2 * \log(e, -1 + x) + 2 * \log(e, 1 + x) = 3$

2) $2 * \log(e, -1 + x) + \log(e, 1 + x) = 3$

3) $\log(e, -1 + x) + 2 * \log(e, 1 + x) = 3$

4) $\log(e, -1 + x) + \log(e, 1 + x) = 4$

[End of set]

> % The problems are all rather similar.
> % Equations 2 and 3 give rise to
> % cubics which LP cannot solve.
> % Here equation 1 is chosen

Which problem do you want run? Type number or n for none.
Problem: *1*.

Equation

$2 * \log(e, -1 + x) + 2 * \log(e, 1 + x) = 3$

selected.

Run begins:

Trying indicated schema step of applying Attraction.

Attracting Equation to obtain

$\log(e, (1 + x) * (-1 + x)) = (3/2)$.

> % Unlike PRESS, LP applies Attraction
> % exhaustively. Here Attraction has
> % applied twice, first 2 was taken out as
> % a common factor and transferred to the
> % rhs, then the log terms have been added.

Trying indicated schema step of applying method(auto4).

Using rule collect_rule to apply new method, auto4, to obtain

$\log(e, x \verb|^| 2 + -1) = (3/2)$.

APPENDIX V

```
Trying indicated schema step of applying Isolation.

Isolating Equation to obtain

x = (1 + e ^ (3/2)) ^ (1/2).

[End of solution]
                    % The schema solves the problem without
                    % difficulty.
Schema solves test problem.

yes
| ?- halt.
                    % End the run
```

REFERENCES

[1] Aiello, L. and Weyhrauch, R.W. Using Meta-Theoretic Reasoning to do Algebra. In Bibel, W. and Kowalski, R. (editors), *5th Conference on Automated Deduction*, pages 1-13. Springer Verlag, 1980. Lecture Notes in Computer Science No. 87.

[2] Allen, J.F. and Perrault, C.R. Analyzing Intentions in Utterances. *Artificial Intelligence* 15:143-178, 1980.

[3] Aubin, R. Some generalization heuristics in proofs by induction. In Huet, G. and Kahn, G. (editors), *Actes du Colloque Construction: Amelioration et verification de Programmes*. Institut de recherche d'informatique et d'automatique, 1975.

[4] Bowen K.A. and Kowalski R.A. *Amalgamating Language and Metalanguage in Logic Programming*. Research Report, School of Computer and Information Science, Syracuse University, June, 1981.

[5] Bowen, D.L., Byrd, L., Pereira, F.C.N., Pereira, L.M. and Warren, D.H.D. *Decsystem-10 Prolog User's Manual*. Occasional Paper 27, Dept. of Artificial Intelligence, Edinburgh, November, 1982.

[6] Boyer, R.S. and Moore J.S. Proving theorems about LISP functions. In Nilsson, N. (editor), *Proceedings of the third IJCAI*, pages 486-493. International Joint Conference on Artificial Intelligence, August, 1973. Also available from Edinburgh as DCL memo No. 60.

[7] Boyer, R.S. and Moore J.S. *A Computational Logic*. Academic Press, 1979. ACM monograph series.

[8] Brazdil, P. *A model for error detection and correction*. PhD thesis, University of Edinburgh, 1981.

[9] Bundy, A. *Analysing Mathematical Proofs (or reading between the lines)*. Research Report 2, Dept. of Artificial Intelligence, Edinburgh, May, 1975. A shorter version is in the proceedings of the fourth IJCAI.

[10] Bundy, A. *An elementary treatise on equation solving*. Working Paper 51, Dept. of Artificial Intelligence, Edinburgh, 1979.

[11] Bundy, A. *Using Winston to Learn Equation Solving Strategies*. Internal Note 76, Dept. of Artificial Intelligence, Edinburgh, March, 1981.

[12] Bundy, A. *Using Focussing to learn isolation rules*. Internal Note 120, Dept. of Artificial Intelligence, Edinburgh, July, 1982.

[13] Bundy, A. *The Computer Modelling of Mathematical Reasoning*. Academic Press, 1983.

[14] Bundy, A. and Silver, B. Homogenization: Preparing Equations for Change of Unknown. In Schank, R. (editor), *Proceedings of IJCAI-81*. International Joint Conference on Artificial Intelligence, 1981. Longer version available from Edinburgh as DAI Research Paper No. 159.

REFERENCES

[15] Bundy, A. and Silver, B. A critical survey of rule learning programs. In Raulefs, P. (editor), *Proceedings of ECAI-82*, pages 150-157. European Conference on Artificial Intelligence, 1982. Also avaliable from Edinburgh as DAI Research Paper No. 169.

[16] Bundy, A. and Sterling L.S. *Meta-level Inference in Algebra*. Research Paper 164, Dept. of Artificial Intelligence, Edinburgh, September, 1981. Presented at the workshop on logic programming for intelligent systems, Los Angeles, 1981.

[17] Bundy, A., Byrd, L., Luger, G., Mellish, C., Milne, R. & Palmer, M. *Mecho: A program to solve Mechanics problems*. Working Paper 50, Dept. of Artificial Intelligence, Edinburgh, 1979.

[18] Bundy, A., Byrd, L., Luger, G., Mellish, C., Milne, R. & Palmer, M. Solving Mechanics Problems Using Meta-Level Inference. In Buchanan, B.G. (editor), *Proceedings of IJCAI-79*, pages 1017-1027. International Joint Conference on Artificial Intelligence, 1979. Also available from Edinburgh as DAI Research Paper No. 112.

[19] Bundy, A., Silver, B. and Plummer, D. An Analytical Comparison of some Rule Learning Programs. In *Third Annual Technical Conference of the British Computer Society's Expert Systems Specialist Group*. British Computer Society, 1983.

[20] Bundy, A., Silver, B. and Plummer, D. *An Analytical Comparison of some Rule Learning Programs*. Research Paper, Dept. of Artificial Intelligence, Edinburgh, 1984. Submitted to Artificial Intelligence Journal Earlier versions published in Procs.. of ECAI-82 and Procs. of Third BCS Expert Systems Group.

[21] Bundy, A. & Welham, B. Using meta-level inference for selective application of multiple rewrite rules in algebraic manipulation. *Artificial Intelligence* 16(2):189-212, 1981.

[22] Carry L.R., Lewis, C. and Bernard, J. *Psychology of Equation Solving: An information processing study*. Technical Report, Dept. of Curriculum Development, University of Texas at Austin, 1981.

[23] Clocksin, W.F. and Mellish, C.S. *Programming in Prolog*. Springer Verlag, 1981.

[24] Conte, S.D. and de Boor, C. *Elementary Numerical Analysis*. McGraw-Hill Kogakusha, 1972.

[25] Cotton, J.W. Personal communication. 1977.Letter describing plans to include instruction in PRESS processes in remedial Mathematics CAI system.

[26] Cotton, J., Byrd, L. and Bundy, A. *How can Algebra steps be learned by students with only arithmetic skills*. Working Paper 84, Dept. of Artificial Intelligence, Edinburgh, 1981.

[27] Davis, R. Meta-Rules: Reasoning about Control. *Artificial Intelligence* 15:179-222, 1980.

[28] Davis, R. and Buchanan, B.G. Meta-level knowledge: overview and applications. In Reddy, R. (editor), *Proceedings of IJCAI-77*, pages 920-927. IJCAI, 1977.

[29] Davis, R., Buchanan, B.G. and Shortliffe, E.H. Production rules as a representation for a knowledge-based consultation program. *Artificial Intelligence* 8(1), February, 1977.

[30] DeJong, G. An approach to Learning From Observation. In Michalski, R.S (editor), *Proceedings of the International Machine Learning Workshop*, pages 171-176. University of Illinois, June, 1983.

[31] DeJong, G. Acquiring Schemata Through Understanding and Generalizing Plans. In Bundy, A. (editor), *Proceedings of the Eighth IJCAI*, pages 462-464. International Joint Conference on Artificial Intelligence, 1983.

[32]   Digricoli, V.J. Resolution by Unification and Equality. In Joyner, W.H. (editor), *Proceedings of the Fourth Workshop on Automated Deduction*, pages 43-52. Feb, 1979.

[33]   Ernst, G. and Newell, A. *GPS: A Case Study in Generality and Problem Solving*. Academic Press, 1969.

[34]   Evans, T.G. A Program for the Solution of Geometric-Analogy Intelligence Test Questions. In Minsky, M.L. (editor), *Semantic Information Processing*, pages 271-353. MIT Press, Cambridge, Mass., 1968.

[35]   Fikes, R.E., and Nilsson, N.J. STRIPS: A new approach to the application of theorem proving to problem solving. *Artificial Intelligence* 2:189-208, 1971.

[36]   Fikes, R.E., Hart, P.E. and Nilsson, N.J. Learning and executing generalized robot plans. *Artificial Intelligence* 3:251-288, 1972.

[37]   Hayes, P. Computation and deduction. In *Proc. of MFCS Symposium*. Czech. Academy of Sciences, 1973.

[38]   Hayes, P. In defence of logic. In *Proceedings of IJCAI-77*. International Joint Conference on Artificial Intelligence, 1977.

[39]   Hearn, A.C. REDUCE: A user-oriented interactive system for Algebraic simplification. In *Interactive systems for experimental Applied Mathematics*, pages 79-90. Academic Press, New York, 1967.

[40]   Huet, G. *Confluent reductions: Abstract properties and applications to term rewriting systems*. Rapport de Recherche 250, Laboratoire de Recherche en Informatique et Automatique, IRIA, France, August, 1977.

[41]   Knuth, D.E. and Bendix, P.B. Simple word problems in universal algebra. In Leech, J. (editor), *Computational problems in abstract algebra*, pages 263-297. Pergamon Press, 1970.

[42]   Kowalski, R. *Artificial Intelligence Series*: *Logic for Problem Solving*. North Holland, 1979.

[43]   Laird, J.E., Rosenbloom, P.S. and Newell, A. Towards Chunking as a General Learning Mechanism. In Brachman, R. (editor), *Procs. of AAAI-84*, pages 188-192. American Association for Artificial Intelligence, 1984.

[44]   Langley, P. Strategy Acquisition Governed by Experimentation. In Raulefs, P. (editor), *Proceedings of ECAI-82*, pages 171-176. European Conference on Artificial Intelligence, 1982.

[45]   Langley, P. Learning Effective Search Heuristics. In Bundy, A. (editor), *Proceedings of the Eighth IJCAI*, pages 419-421. International Joint Conference on Artificial Intelligence, 1983.

[46]   Lenat, D.B. Theory Formation by Heuristic Search. The Nature of Heuristics II: Background and Examples. *Artificial Intelligence* 21(1-2):31-60, March, 1983.

[47]   Lenat, D.B. EURISKO: A Program That Learns New Heuristics and Domain Concepts. The Nature of Heuristics III: Program Design and Results. *Artificial Intelligence* 21(1-2):61-98, March, 1983.

[48]   Lenat, D.B. The Role of Heuristics in Learning by Discovery: Three Case Studies. In Michalski, R.S, Carbonell, J.F. and Mitchell, T.M. (editors), *Machine Learning*, pages 243-306. Tioga Press, 1983.

[49]   Mathlab Group. *MACSYMA Reference Manual*. Technical Report, MIT, 1977.

[50]   McCarthy, J. Programs with Common Sense. In *Mechanisation of Thought Processes (Proceedings of a symposium held at the National Physics Laboratory, London, Nov 1959)*, pages 77-84. HMSO, London, 1959.

REFERENCES

[51]   Minton, S. Constraint-Based Generalization: Learning Game-Playing Plans From Single Examples. In Brachman, R. (editor), *Procs. of AAAI-84*, pages 251-254. American Association for Artificial Intelligence, 1984.

[52]   Mitchell, T.M. *Version Spaces: An approach to concept learning*. PhD thesis, Stanford University, 1978.

[53]   Mitchell, T.M. *Toward Combining Empirical and Analytical Methods For Inferring Heuristics*. Technical Report LCSR-TR-27, Laboratory for Computer Science Research, Rutgers University, 1982.

[54]   Mitchell, T.M. Learning and Problem Solving. In Bundy, A. (editor), *Proceedings of the Eighth IJCAI*, pages 1139-1151. International Joint Conference on Artificial Intelligence, 1983.

[55]   Mitchell, T.M., Utgoff, P. E., Nudel, B. and Banerji, R. Learning problem-solving heuristics through practice. In *Proceedings of IJCAI-81*, pages 127-134. International Joint Conference on Artificial Intelligence, 1981.

[56]   Mitchell, T.M., Utgoff, P. E. and Banerji, R. Learning by Experimentation: Acquiring and modifying problem-solving heuristics. In Michalski, R.S, Carbonell, J.F. and Mitchell, T.M. (editors), *Machine Learning*, pages 163-190. Tioga Press, 1983.

[57]   Mostow, D.J. *Mechanical transformation of task heuristics into operational procedures*. PhD thesis, Carnegie-Mellon University, 1981.

[58]   Neves, D.M. A computer program that learns algebraic procedures by examining examples and working problems in a textbook. In *Proceedings of the Second National Conference*, pages 191-195. Canadian Society for Computational studies of Intelligence, 1978.

[59]   Neves, D.M. *Learning Procedures from Examples*. PhD thesis, Dept. of Psychology, Carnegie-Mellon University, 1981.

[60]   Nilsson, N.J. *Principles of Artificial Intelligence*. Tioga Pub. Co., Palo Alto, California, 1980.

[61]   O'Keefe, R.A. *Automated Statistical Analysis*. Working Paper 104, Dept. of Artificial Intelligence, Edinburgh, 1982.

[62]   O'Rorke, P. Generalization for Explanation-Based Schema Acquisition. In Brachman, R. (editor), *Procs. of AAAI-84*, pages 260-263. American Association for Artificial Intelligence, 1984.

[63]   Sacerdoti, E.D. Planning in a heirarchy of abstraction spaces. *Artificial Intelligence* 5:115-135, 1974.

[64]   Sacerdoti, E.D. *Artificial Intelligence Series: A Structure for Plans and Behaviour*. North Holland, 1977.

[65]   Silver, B. The application of Homogenization to simultaneous equations. In Loveland, D.W. (editor), *6th Conference on Automated Deduction*, pages 132-143. Springer Verlag, 1982. Lecture Notes in Computer Science No. 138. Also available from Edinburgh as Research Paper 166.

[66]   Silver, B. *Learning Algebraic Methods from Examples - A Progress Report*. Working Paper 129, Dept. of Artificial Intelligence, Edinburgh, November, 1982.

[67]   Silver, B. Learning Equation Solving Methods from Worked Examples. In Michalski, R.S (editor), *Proceedings of the International Machine Learning Workshop*, pages 99-104. University of Illinois, June, 1983. Also available from Edinburgh as Research Paper 188.

REFERENCES

[68] Silver, B. Learning Equation Solving Methods from Examples. In Bundy, A. (editor), *Proceedings of the Eighth IJCAI*, pages 429-431. International Joint Conference on Artificial Intelligence, 1983. Also available from Edinburgh as Research Paper 184.

[69] Silver, B. *Using Meta-Level Inference To Constrain Search And To Learn Strategies In Equation Solving*. PhD thesis, Dept. of Artificial Intelligence, Edinburgh, 1984.

[70] Silver, B. Precondition Analysis: Learning Equation Solving Strategies from Worked Examples. In Michalski, R.S., Carbonell, J.G. and Mitchell, T.M. (editors), *Machine Learning 2*. Morgan-Kaufmann, 1985.

[71] Skinner, D. *A computer program to perform integration by parts*. Working Paper 103, Dept. of Artificial Intelligence, Edinburgh, 1981.

[72] Smith, R.G., Mitchell, T.M., Chestek, R.A., and Buchanan, B.G. A model for Learning Systems. In Reddy, R. (editor), *Proceedings of IJCAI-77*, pages 338-343. International Joint Conference on Artificial Intelligence, 1977.

[73] Stefik, M. Planning and Meta-planning (MOLGEN: Part 2). *Artificial Intelligence* 16:141-170, 1981.

[74] Stefik, M. Planning with Constraints (MOLGEN: Part 1). *Artificial Intelligence* 16:111-140, 1981.

[75] Sterling, L. and Bundy, A. Meta-level Inference and Program Verification. In Loveland, D.W. (editor), *6th Conference on Automated Deduction*, pages 144-150. Springer Verlag, 1982. Lecture Notes in Computer Science No. 138. Also available from Edinburgh as Research Paper 168.

[76] Sterling, L., Bundy, A., Byrd, L., O'Keefe, R., and Silver, B. Solving Symbolic Equations with PRESS. In Calmet, J. (editor), *Computer Algebra, Lecture Notes in Computer Science No. 144.*, pages 109-116. Springer Verlag, 1982. Longer version available from Edinburgh as Research Paper 171.

[77] Sussman G.J. *Artificial Intelligence Series: A Computer Model of Skill Acquisition*. North Holland, 1975.

[78] Tate, A. *Project Planning Using a Hierarchic Non-Linear Planner*. Research Report 25, Dept. of Artificial Intelligence, Edinburgh, 1976.

[79] Utgoff, P.E. Adjusting Bias In Concept Learning. In Michalski, R.S (editor), *Proceedings of the International Machine Learning Workshop*, pages 105-109. University of Illinois, June, 1983.

[80] Utgoff, P.E. Adjusting Bias In Concept Learning. In Bundy, A. (editor), *Proceedings of the Eighth IJCAI*, pages 447-449. International Joint Conference on Artificial Intelligence, 1983.

[81] Wallen, L.A. *Using Proof Plans to Control Deduction*. Research Paper 185, Dept. of Artificial Intelligence, Edinburgh, February, 1983.

[82] Warren, D.H.D.. Prolog on the DECsystem-10. In Michie D (editor), *Expert Systems in the Micro-Electronic Age*, pages 112-121. Edinburgh University Press, 1979.

[83] Welham, R and Bundy, A. *Equation solving: A progress report*. Working Paper 29, Dept. of Artificial Intelligence, Edinburgh, June, 1978.

[84] Weyhrauch, R.W. Prolegomena to a theory of mechanized formal reasoning. *Artificial Intelligence* 13:133-170, 1980.

[85] Weyhrauch, R.W. An example of FOL using metatheory: formalizing reasoning systems and introducing derived inference rules. In Loveland, D.W. (editor), *6th Conference on Automated Deduction*, pages 151-158. Springer-Verlag, 1982.

[86] Wilensky, R. *Understanding Goal Based Stories*. PhD thesis, Yale Universtiy, 1978.

[87] Winograd, T. *Understanding Natural Language*. Edinburgh University Press, 1972.

[88] Winston, P. Learning structural descriptions from examples. In Winston, P.H. (editor), *The psychology of computer vision*. McGraw Hill, 1975.

[89] Young, R.M., Plotkin, G.D. and Linz, R.F. Analysis of an extended concept-learning task. In Reddy, R. (editor), *Proceedings of IJCAI-77*, pages 285. International Joint Conference on Artificial Intelligence, 1977.

INDEX

A level **6**, 7
ABSTRIPS 4, 139
Add list **4, 76**, 135
ALEX 4, 75, 76, 83, 118, **131**, 151
Algebraic equations **42**
Argument-Reduction 163, **165**, 166, 171
Arithmetic Progression Specialist **27**, 112
Attraction **21**, 80, 84, 91, 108, 109, 112, 113, 123
Aubin 64

Back propagation **147**
Back-substitution 54, **55**
Backtracking 30, 106
Base class **42**, 52
Base ten method **181**
Basic Method **22**
Boyer 64
Boyer-Moore theorem-prover 63
Buchanan 129
Bundy 3, 41, 54, 117

Change of Unknown 41, 42, 43, 57, 75, 78, 79, 86, 92, 99, 101, 102, 107, 109, 114, 123, 124, 125
Change of Unknown Method **25**
Changed variable equation **25**
Closeness **18**
Collection **19**, 49, 60, 79, 80, 81, 91, 92, 108, 109, 112, 122, 169, 171
Combinatorial explosion **7**
Common subterm **26**
Concept-learning 115
Current indicated method **98**
Current schema section **100**

Davis 129, 151
DeJong 6, 119
Delete list **4, 76**, 135
Depth **17**

Effects **4, 76**
Elementary Method **54**, 58
Elimination 54, **58**
ELM 118
EURISKO 130, 131
Explanation tree **144**
Explanatory Schema Acquisition 6, **119**
Exponential and Hyperbolic Offenders Set **182**
Exponential Offenders Set **177**
Expression tree 17
Extended Homogenization 41, 54, **56**

Factorization **22**, 71, 72, 75, 76, 78, 79, 86, 88, 92, 93, 98, 101, 102, 107, 108, 113, 124, 170, 171

INDEX

Factorization Preparation **26**, 71, 79, 81, 86, 87, 89, 92, 107, 109, 113, 170
Fikes 135
Flow-chart 99
FOL 128, 131, 151
FOO 140, 143
Function Stripping 160, **160**, 169, 170, 171
Function Swapping **166**, 166
Function-Collection **167**
Function-Isolation 163, **163**, 165, 166

Generalization 63
Generalization language **143**
Generating task **93**
Given task 97
Goal Directed Learning 6
Goal-directed learning **142**
GOLUX 128, 129
GPS 8, 135

Hayes 128
Heuristic waterfall **29**
Homogenization 26, 31, **41**, 77, 80, 113, 114, 117, 177
Hyperbolic Offenders Set **181**

Identities **173**
IMPRESS 10, 34
Input equation **42**, 49
Inverse-Cancellation 163, **164**, 165, 166, 171
Isolation **19**, 71, 77, 78, 86, 107, 109, 112, 114, 119, 122, 160, 170, 171

Kernel **137**
Key method 78
Key methods 71, **78**

Learning by doing **111**
Learning by rote **98**
Least covering tree **17**
Least-dominating **20**
Lenat 129
LEX 4, 76, 111, 118, **141**
LEX2 **140**, 183
Logarithmic 160
Logarithmic Method 114, **172**
Logarithmic Offenders Set **179**
LP methods 76

MACROPS 135, 138
MACSYMA 9, **37**
Major aim **72**
Major steps **67**
McCarthy 2
MECHO **13**
Meta-knowledge 127
Meta-level Inference **2**, 15, 32, 39
Meta-meta-level 34
Meta-theory 2, **32**
Method 16
Methods 5
Mitchell 6, 140, 184
Mixed offenders set 46
MOLGEN 127
Moore 64
Mostow 128

MYCIN   129

Nasty Function Methods   78, 159
Neves   68, 75, 83
Not immediately parsed   81

Object-level   3, 15
Offenders set   42
Offending terms   42
Operationalization   128
Operator Identification   67, 81, 84
Output equation   42, 49

Plan Recognition   96
PLANEX   99, **136**, 138
Polynomial Methods   23, 78, 107
Position   17
Postconditions   5, **76**, 77
Power method   180
Precondition Analysis   1, 4
Precondition Analysis phase   67, 85
Primary methods   29
Priority ordering   36
Problem generator   183
Proof checking   84
Psychological validity   38

Quotient Removal   160, **161**

R Elementary   14
Rational functions   **161**
Recognizer   36
Recognizers   16
REDUCE   9, **37**
Reduced term   **42**, 44, 177
Rewrite rule   44
Rewrite rules   **14**, 79
Rule list   90

SAGE   76, 111, 118
SAGE,   4
Schema   **72**, 93
Schema method   72
Schema methods   93
SCOPE   10, 84
Search space   34
Secondary methods   29
Self-improving algebra system   10
SIMPLE   83
Simplicity method   46, 49
Simplicity metric   46
Simultaneous equations   54
Size   17
Solution to the equation   13
Spectrum of generalization strategies   140
STRIPS   4, 76, 99, 118, **135**, 140, 151
Substitution equation(s)   25
Support   137

Table of Conditions   80, 102, 103, 107, 143
Table of Connections   80, 89, 103, 107
TEIRESIAS   129, 131, 151
Triangle table   136

Trigonometric Factorization   112, 124, 159, **167**
Trigonometric offenders set   46

Unification   62
Utgoff   **148**

Waterfall   **29**, 38, 105
Welham   3
Wenger   39
Weyhrauch   128
Worked examples   74

X-offenders set   57
X-reduced term   57
X_j-offenders set   57
X_j-reduced term   57